Patterns of Progress

CONTINUOUS CULTURE IN MICROBIAL PHYSIOLOGY AND ECOLOGY

Patterns of Progress

General Editor: J.Gordon Cook, B.Sc., Ph.D., F.R.I.C.

Concise, authoritative reviews of technical topics

Microbiology Series

Bacterial Conjugation, D.J. Finnegan
The Microflora of the Rumen, P.N. Hobson
The Mycobacteria, C. Ratledge
Scrapie in the Mouse, R.H. Kimberlin
Microbial Response to Mild Stress, R.E. Strange
Modern Views on Microbial Pathogenicity, H. Smith
Epidemiology and Infections, C.E. Gordon Smith
Yield Studies in Microorganisms, A.H. Stouthamer
Bacterial Reaction to Radiation, B.A. Bridges
Bacterial Transduction, S.B. Primrose
The Oxygen Metabolism of Microbes, D.E.F.Harrison
Continuous Culture in Microbial Physiology and Ecology, H. Veldkamp

CONTINUOUS CULTURE IN MICROBIAL PHYSIOLOGY AND ECOLOGY

H. Veldkamp
Professor of Microbiology, University of Groningen,
Biological Centre, Haren (Gr.), The Netherlands.

MEADOWFIELD

Meadowfield Press Ltd.,
I.S.A. Building, Shildon,
Co. Durham, England.

ISBN 0 904095 25 8

Printed in Great Britain at
Upton Printing Services,
Bushey, England

CONTENTS

PREFACE

The purpose of this book is to survey some of the possibilities which the continuous culture technique presents in studying the interrelations between microbes and their biotic and abiotic environments. The choice of experiments described was an arbitrary one. For a more detailed discussion of their results the reader is referred to the original literature. I am indebted to Dr W. Harder for valuable discussions during the preparation of the manuscript.

H. Veldkamp

1

INTRODUCTION

In a continuous culture fresh growth medium is introduced continuously into the culture vessel, culture liquid containing microorganisms being concomitantly removed at the same rate. The unique properties of continuous cultures make possible the study of physiological and ecological problems which cannot be tackled with any type of closed culture system.

The most commonly used continuous culture is a single stage flow-controlled system. This was first described by Monod (1950), and independently by Novick and Szilard (1950), who named it the "chemostat". This name is still in use, although it is also applied, and perhaps more properly so, for systems in which the concentration of a nutrient is held at a fixed value and the rate of medium supply adjusts itself.

1.1 *Flow-controlled Continuous Cultures*

Single-stage Chemostat

Technical details about chemostats have been given by several authors, e.g. Herbert *et al.* (1965), Řičica (1966), Baker (1968), Evans *et al.* (1970), Solomons (1972), Veldkamp and Kuenen (1973), Harder *et al.* (1974).

The first mathematical analyses of the nutrient-limited continuous culture were given by Monod (1950), and Novick and Szilard (1950). Concise descriptions of basic principles were given subsequently by Herbert *et al.* (1956), Powell (1965) and Kubitschek (1970). More elaborate discussions of chemostat theory are to be found in the publications of Moser (1958), Fencl (1966), Powell (1967), Tempest (1970), and Pirt (1975).

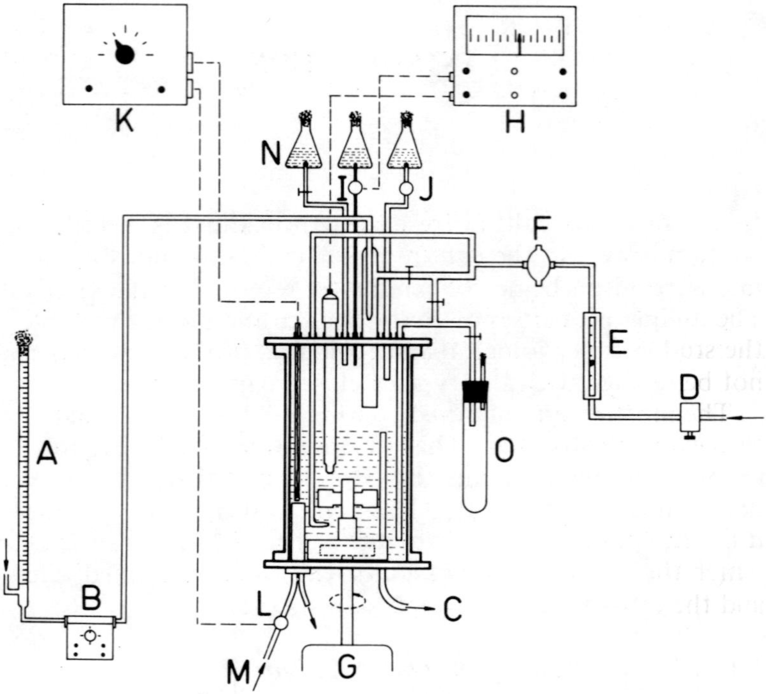

Fig.1. Schematic diagram of single-stage chemostat. Details of fermentor described by Harder *et al*., (1974).
A: burette (connected to medium reservoir) for measuring medium flow rate; B: medium pump; C: outlet for medium and air; D: gas flow controller; E: gas flow meter; F: filter; G: AC motor of magnetic drive; H: pH controller; I: Peristaltic pump or magnetic valve connected to inlet tube for acid or alkali; J: Peristaltic pump for antifoam addition; K: temperature controller; L: circulation pump; M: connection to temperature controlled water bath; N: inoculation flask; O: sampling flask.

A chemostat with commonly used ancillary equipment is shown diagrammatically in Fig.1. Fresh sterile medium is pumped from a medium reservoir into the culture vessel from which culture liquid is removed at the same rate (F). The culture volume (V) is kept constant, and the dilution rate (D) is defined as D = F/V (dimension: h^{-1}). The key to the mode of action of the chemostat is the composition of the medium in the reservoir. All components in this medium, other than the growth-limiting substrate, are present in excess. The symbol S_R represents the concentration of the growth-limiting substance in the reservoir; the symbol s represents its concentration in the culture vessel. The specific growth rate (μ) of a bacterial culture is defined as:

$$\mu = \frac{1}{x} \cdot \frac{dx}{dt} = \frac{d \ln x}{dt} = \frac{\ln 2}{t_d} \qquad (1)$$

where x is organism concentration (mg dry weight/ml) at time t, and t_d is time required for the concentration of organisms to double. The specific growth rate μ depends on s. This dependence is empirically described by the Monod (1942, 1950) equation:

$$\mu = \mu_{max} \left(\frac{s}{K_s + s} \right) \qquad (2)$$

where K_s is a saturation constant, numerically equal to the concentration of the growth-limiting substrate at $0.5\mu_{max}$.

When a chemostat is inoculated, and the dilution rate does not exceed a certain critical value D_c:

$$D_c = \mu_{max} \left(\frac{S_R}{K_s + S_R} \right) \qquad (3)$$

the change in organism concentration (x) is given by

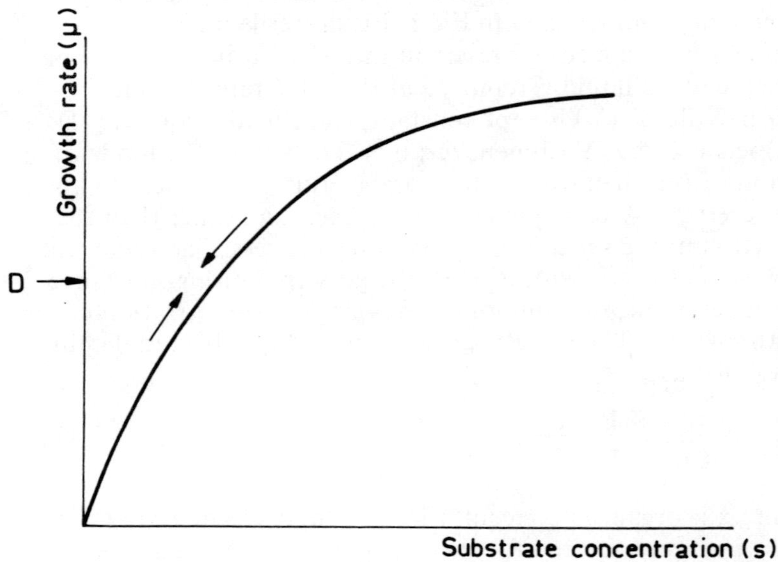

Fig.2. Specific growth rate (μ) as a function of substrate concentration (s). When the dilution rate (D) is fixed at a certain value and the culture vessel is inoculated, a steady state is automatically reached in which $\mu = D$. This holds for conditions in which initially either $\mu > D$, or $\mu < D$. For instance: when initially $\mu > D$, the population density (x) increases, and therefore the concentration of the limiting substrate (s) decreases, and thus μ decreases until $\mu = D$.

$$\frac{dx}{dt} = \text{growth - output}$$

or,

$$\frac{dx}{dt} = \mu x - Dx \qquad (4)$$

When the dilution rate D is set at a fixed value, x and s automatically reach a steady state level in which

$$\frac{dx}{dt} = \frac{ds}{dt} = 0. \text{ (cf. Fig.2).}$$

The change in substrate concentration (s) is given by:

$$\frac{ds}{dt} = \text{input - output - consumption.}$$

The relation between rate of growth and substrate consumption was described by Monod (1942) as

$$\frac{dx}{dt} = - Y \frac{ds}{dt} \qquad (5)$$

The change in s therefore is:

$$\frac{ds}{dt} = DS_R - Ds - \frac{\mu x}{Y} \qquad (6)$$

From equation (4) and (6) the steady state values of organism concentration ($\bar{x}$) and growth-limiting substrate ($\bar{s}$) can be calculated:

$$\bar{x} = Y (S_R - \bar{s}) \qquad (7)$$

$$\bar{s} = K_s \left(\frac{D}{\mu_{max} - D}\right) \qquad (8)$$

These equations have the following constants: S_R, Y, K_s, and μ_{max}. Once their values are known, the steady state values $\bar{x}$ and $\bar{s}$ can be predicted for any dilution rate (cf. Fig.3).

The above considerations show that the chemostat offers possibilities which are absent in any closed culture system. During growth in batch culture microbes continuously change their environment. Unicellular organisms thriving in continuously changing natural environments are adapted to give quick responses to these changes and therefore the morphological and metabolic properties of cells grown in batch culture are apt to change during the growth period. This can of course be minimized by frequent transfers to fresh medium. However, cells cultivated in such a way grow at maximal rate. In the chemostat cells grow at submaximal rate, as they usually do in their natural environment. And these cells can be grown in steady states at any of a whole range of growth rates. The only factor determining a particular growth rate is the concentration of the growth-rate-limiting substrate, which is determined by the dilution rate (cf. eq. 8). In addition, any substrate that the organism needs can be made growth-limiting. Thus the nutritional status of an organism as well as its growth rate can be varied at will. In other words, a pure culture of an organism can be grown in steady state in the chemostat under a multitude of growth conditions, each set of conditions giving a phenotypically unique type of cells. Once a steady state is reached neither the properties of the culture, nor those of the environment undergo further change. The culture then has become time-independent, and conditions are under full control. Therefore, experiments with the chemostat are highly reproducible.

Ecologists should realize that although a microbial culture in the chemostat grows under substrate limitation, which undoubtedly is the rule in nature, the chemostat does not mimic a natural environment in any way. In natural environments growth conditions change continuously, and in evolution selec-

Fig.3. Steady-state relationships in a continuous culture (theoretical). The steady-state values of substrate concentration, bacterial concentration and output at different dilution rates were calculated from equations (7) and (8), for an organism with the following growth constants: $\mu_m = 1.0\ hr^{-1}$, $Y = 0.5$ and $K_s = 0.2$ g/l; substrate concentration in the inflowing medium, $S_R = 10$ g/l (Herbert *et al.*, 1956).

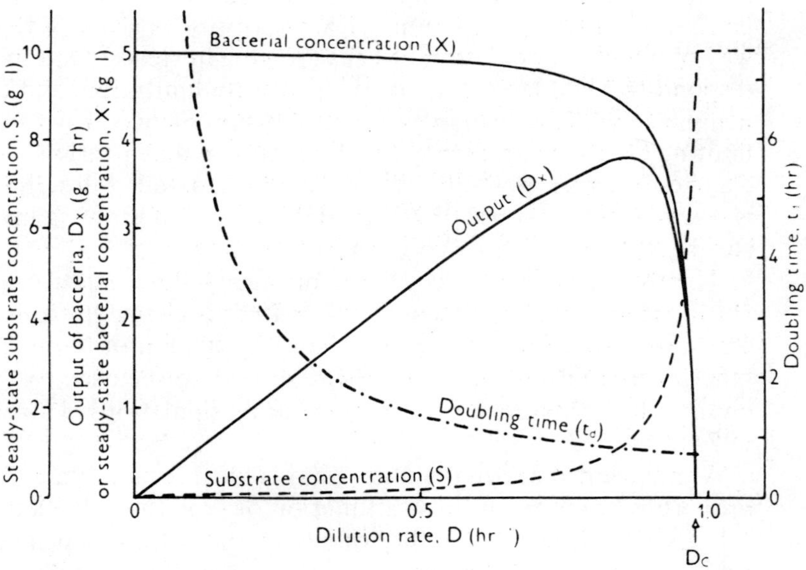

Bacterial concentration (X)
Output (Dx)
Doubling time (td)
Substrate concentration (S)
Steady-state substrate concentration, S, (g l)
Output of bacteria, Dx (g l hr) or steady-state bacterial concentration, X, (g l)
Doubling time, t, (hr)
Dilution rate, D (hr)
Dc
0
2
4
6
8
10
1
3
5
0.5
1.0

Fig.3

tion favoured those organisms that could cope with these changes, rather than with steady state conditions as occur in the chemostat and rarely in nature.

The chemostat is particularly useful as a research tool because it makes possible the study of the responses an organism can give at the metabolic and population level to controlled changes in the environment. Examples of experiments of this type are given in Chapter 4; in section 4.1 a description is given of the way in which a population of cells growing in steady state may respond to a sudden release of the substrate limitation, and the manner in which simultaneous consumption of two growth-limiting carbon sources may be affected by concentration changes of these substrates. In section 4.2 a detailed description is given of the response of a bacterial population growing in steady state to sudden changes in temperature.

The theory given above is based on steady state kinetics. During transient state, when a pure culture is changing from one steady state to another, the Michaelis-Menten or Monod kinetics are seldom applicable. They should not be used to predict the behaviour of complex natural populations (Jannasch, 1974).

As mentioned above the chemostat is ideal for studying the properties of an organism as a function of growth rate. Section 2.1 describes how the viability of a bacterial culture depends on growth rate, and the profound effect that the nutritional status (kind of growth-limitation) may have on viability.

Taxonomists rarely describe the morphology of a microbial species as a function of growth rate. Perhaps they should, as cell appearance may dramatically change with growth rate, as is described in section 5.1

The chemostat is highly selective. When a mixture of organisms is competing for a single growth-limiting substrate (and no other interactions occur) one organism will reach steady state conditions, whereas the slower growing ones are washed out. This may happen in a pure culture in which a mutant arises that grows faster than the parent strain under the given conditions.

An interesting possibility offered by the chemostat is to study competition at extremely low concentrations of growth-limiting substrates, such as occur in natural environments. Studies of this type have shown that in soil and natural waters organisms occur that grow relatively fast at very low concentration of growth-limiting substrates. These organisms grow relatively slowly at higher concentrations of these substrates and therefore do not show up in the usual batch culture enrichments. Experiments of this type are described in section 2.2.

When chemostat theory as described above was tested experimentally, some interesting deviations were found. The yield value (Y), which on the basis of batch culture data was supposed to be a constant, in fact varies with growth rate. First evidence came from chemostat studies of C-limited cultures which showed a decreasing cell yield with decreasing growth rates, which was not predicted by theory. It appears that microbial cells need, independent of growth rate, a certain small amount of energy per unit time per mg dry weight for purposes other than growth ("maintenance requirement"). In fast-growing cells exposed to relatively high concentrations of the C- and energy source, this amount of energy is negligibly small as compared to total energy production. However, at very low growth rates, energy consumption for maintenance purposes may become considerable in comparison with the energy used for biosynthesis. Growth yield then is lower than predicted by theory. In ecological terms: at decreasing growth rates, biomass formation decreases, and the accent of the functional status of bacteria in natural ecosystems then shifts to nutrient cycling (cf. Veldkamp, 1975). A detailed discussion of this phenomenon is given in Chapter 3.

Theory also does not take into account that cell composition changes as a function of growth rate. An example is given in section 5.1, showing that the Mg content per mg bacterial dry weight increases with increasing growth rates. Therefore, in Mg-limited cultures, the yield with respect to Mg decreases more with increasing growth rates than predicted by theory.

A population of cells in the chemostat consists of individuals of different age with respect to the division cycle. It is possible to combine continuous culture with cell synchrony (Dawson, 1972). Such cultures are not in steady state.

Multi-stage Systems

The theory of multi-stage continuous cultures has been dealt with by Herbert (1964), Powell and Lowe (1964), Fencl (1966), and Pirt (1975). Herbert (1964) classified multi-stage systems as follows: single stream systems, multi-stream systems, and multi-stage systems with feedback. None of these systems has been used extensively for physiological or ecological studies.

Multi-stage systems allow separation of ecological niches. An example is the work of Abson and Todhunter (1961) who, in an attempt to purify carbonisation effluents, separated in a three-stage system the oxidation of phenols, thiocyanate and thiosulfate, and ammonia.

Another possibility which the multi-stage system offers is the study of predator-prey relations, in which the prey, grown in the first-stage, is fed into a second stage containing the predator. Examples of some of the few studies of this type are given in Chapter 2.

For further reading the following publications, which cover the greater part of multi-stage systems literature, are recommended: Málek and Fencl (1966), Řičica *et al.* (1967), Řičica (1970), Fencl *et al.* (1972). Of particular interest are computer simulations of population dynamics of bacteria and protozoa in activated sludge in single stage (with recycle) and multi-stage systems (Curds, 1971b).

1.2 *Non-flow-controlled Continuous Cultures*

In systems of this type either the concentration of a nutrient is fixed by a probe controlling the medium metering pump (chemostat *sensu stricti*), or the population density is controlled by a photoelectric device (turbidostat).

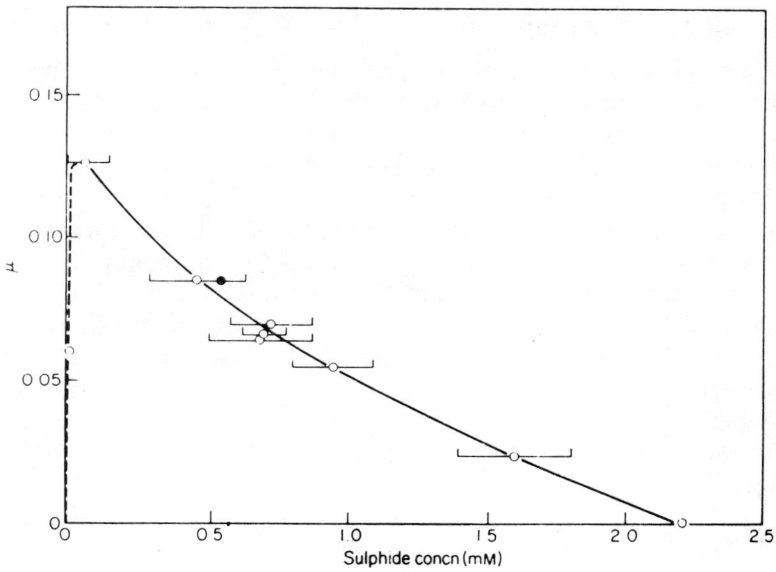

Fig.4. Relation between specific growth rate and sulfide concentration of a *Chromatium* species (van Gemerden and Jannasch, 1971).

The Chemostat *Sensu Stricti*

The following example illustrates why a flow-controlled system cannot always be used. Some substrates become toxic at extremely low concentrations, even for organisms which show an obligate requirement. An example is sulfide, which is needed as an electron donor by a variety of photosynthetic bacteria. As shown in Fig.4, there are two sulfide concentrations for any value of the specific growth rate less than μ_{max}. A stable steady state in a sulfide-limited flow-controlled system is only possible in the sulfide concentration range in which $d\mu/ds > 0$ (Spicer,

1955; Andrews, 1968). The higher sulfide concentration does not allow stable conditions, since small deviations in this concentration result in one of two other steady states: the one at low sulfide concentration (in range $\frac{d\mu}{ds} > 0$), or the one in which not a single cell is left in the system. The problem can be solved by having the medium-flow controlled by a sulfide electrode adjusted to a fixed sulfide concentration. Influx of fresh medium then is only allowed when the sulfide concentration in the culture becomes lower than the set value. In this way a steady-state can be obtained of cells growing at toxic sulfide concentrations, as undoubtedly occurs in natural environments. A chemostat of this type in fact is a "sulfidostat".

A similar arrangement can be devised for any growth-limiting substrate that can thus be controlled. This need not necessarily be restricted to toxic substrate levels. A well-known example is the system in which a feedback-control of the dissolved oxygen concentration is applied in cultivating aerobic microbes (McLennan and Pirt, 1970).

The Turbidostat

As shown in Fig.3, there is in the flow-controlled chemostat a rather steep decrease in organism concentration when the dilution rate reaches its critical value. At these higher dilution rates, a steady state is often difficult to maintain, and growth at maximal rate is not possible in a single stage chemostat. This becomes easy, however, when a turbidostat is used. The operational difference between the two systems is as follows: In a flow-controlled culture the dilution rate is set at a fixed value, and the organism concentration reaches a steady state value. In the turbidostat, the turbidity of the culture (organism concentration) is fixed at a certain value, and when this is exceeded, a medium-metering device comes into operation. In other words, the dilution rate is allowed to adjust itself. As pointed out by Herbert (1958), both systems are in principle essentially similar. They are merely, for practical purposes, complementary.

As many experiments with fast-growing cells can also be

carried out in batch culture, the impact of the turbidostat on the promotion of physiological and ecological knowledge has been small as compared to the flow-controlled chemostat. A review of the present status and future prospects of the turbidostat was given by Watson (1972).

1.3 *Running A Chemostat*

> "All chemostats are continuous fermentors,
> but some are more continuous than others."
> Solomons (1972).

Even though a simple chemostat is easy to run, many little things may disturb its proper functioning. Often these are of a technical nature, and it is impossible to list remedies for all of them. A few general recommendations are listed below.

1. Good mixing is a prerequisite for the proper operation of a chemostat. Effects of incomplete mixing were discussed by Herbert *et al.* (1956), Hansford and Humphrey (1966) and Solomons (1972).
2. When wall-growth occurs, a steady state cannot be attained. The effect of wall-growth is discussed by Topiwala and Hamer (1971), Solomons (1972), Wilkinson and Hamer (1974).
3. Clumping is a source of inhomogeneity. Growth conditions within an aggregation of cells are very different from those outside the clump. In most cases of clumping the only answer is to use another organism.
4. To ascertain whether a particular component in the growth medium is growth-limiting, a shot of a concentrated solution of that component should be injected into the culture through the inoculation port. If the component is growth-limiting, the cell concentration should increase (cf. eq. 7). If a gas (oxygen, methane or nitrogen) is the limiting nutrient, the cell concentration should continue to increase if the flow of culture fluid is stopped for a brief period.
5. The gas blown through cultures should be equilibrated with

water of the same temperature before entering the growth vessel. Otherwise evaporation may be considerable. This may lead to errors in measuring such parameters as yield and maintenance coefficient (cf. King *et al.*, 1972).

6. Theoretically, the size of the inoculum should be of no importance when starting a chemostat culture. In practice, however, the size may be important, particularly if the organism is sensitive to excessive aeration (e.g., a thiobacillus). With strict anaerobes a heavy inoculum can be essential when redox conditions in the vessel are not optimal for growth. An example of an inoculum size effect in mixed culture studies was described by Meers and Tempest (1968).
7. As pointed out by Solomons (1972), dilution rate is based on the volume of liquid in the culture, not on the fermenter contents. However, these are not synonymous. Rapidly stirred cultures may hold up as much as 25% of air in their volume, so the true dilution rate may differ considerably from the apparent value.

2

POPULATION KINETICS

The chemostat has proved to be a useful tool for a variety of population kinetical studies. A few examples dealing with viability, competition for a single growth-limiting substrate, and predator-prey relations are given below. Mixed population studies were reviewed by Bungay and Bungay (1968), Veldkamp and Jannasch (1972), Gaudy and Gaudy (1972), Meers (1973), Jannasch and Mateles (1974), Harrison *et al.* (1976), and Veldkamp (1976). Selection of mutants and genetical aspects of continuous culture studies were dealt with by Kubitschek (1970). A mathematical description of selection of mutants was given by Powell (1958).

2.1 *Viability As A Function Of Growth-rate*

Most studies on the viability of chemostat cultures were carried out with *Aerobacter aerogenes*. With decreasing dilution rates, a decrease in viability was observed (Postgate and Hunter, 1962; Tempest *et al.*, 1967; Sinclair and Topiwala, 1970; Postgate, 1973). The following equation (Sinclair and Topiwala, 1970) gives a fair approximation of the experimental data obtained thus far:

$$\text{viability } v = \frac{\bar{x}}{(\bar{x} + \bar{x}')} = \frac{\bar{x}}{\bar{x}_t} = \frac{D}{D + \gamma} \qquad (9)$$

where $\bar{x}$, $\bar{x}'$ and $\bar{x}_t$ are steady state concentrations of viable cells, non-viable cells, and total cell mass (gm/litre), respectively. D is dilution rate (h^{-1}) and γ is the cell death constant (h^{-1}). Equation (9) can be rearranged:

$$\frac{1}{v} = 1 + \frac{1}{D} \cdot \gamma \qquad (10)$$

When 1/v is plotted against 1/D, a straight line should be obtained of slope γ. Sinclair and Topiwala (1970) thus calculated γ values from their own experimental results, as well as from those obtained by Postgate and Hunter (1962) and Tempest *et al.* (1967). The predicted curves (eq. 9) fit the data quite well. An example is shown in Fig.5.

In *A. aerogenes* the value of γ is dependent on a number of factors, listed by Postgate (1973). It appears that death rate depends on the kind of substrate limitation. A remarkably high viability was found in slow-growing populations limited by ammonia-N, and phosphate. At growth rates below $0.25\mu_{max}$, C-limited cells showed a marked decrease in viability with decreasing growth rate. It may well be that lack of energy for maintenance purposes (cf. Chapter 3) is responsible for the low viability of cultures of C-limited chemoorganotrophs when the C- and energy supply is highly restricted. Mg^{2+}-limited cultures of *A. aerogenes* were very fragile in the sense that viability starts to decrease at relatively high dilution rates (around $0.5\mu_{max}$) and reaches very low values when the dilution rate is progressively decreased. In this respect it is of interest that Strange and Hunter (1967) observed a favourable effect of Mg^{2+} on starving bacterial populations. Evidence was obtained that this was not due only to the stabilizing effect on ribosomes.

Postgate and Hunter (1963) further showed by varying the Mg^{2+} concentration in the inflowing medium at constant dilution rate (cf. eq. 7) that viability was dependent on population density. The proportion of dead organisms was significantly higher at low organism concentrations. A Mg-limited culture should not be considered as being exposed to Mg starvation. Starvation conditions are obtained when no growth-limiting substrate becomes available to the culture. Obviously, the behaviour of such cultures cannot be studied in the chemostat (cf. Postgate, 1967).

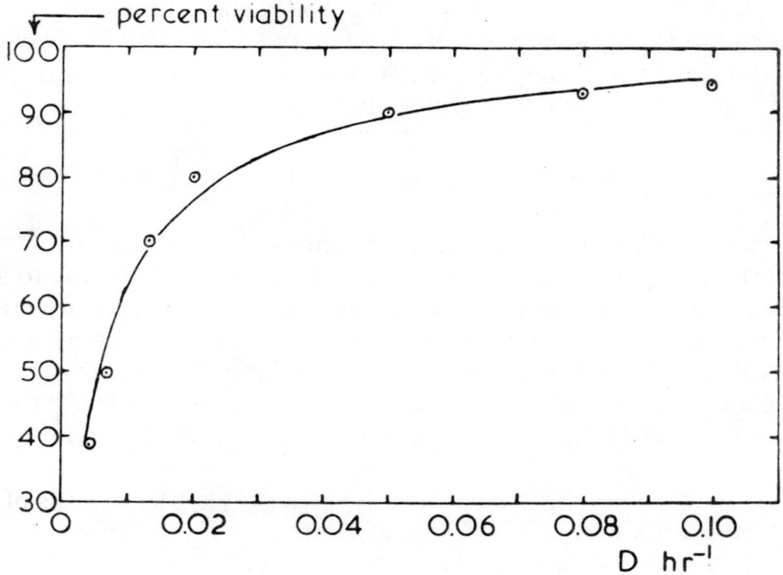

Fig.5. Steady-state viability versus dilution rate as observed in glycerol-limited *Aerobacter aerogenes* (Tempest *et al.*, 1967). Parameter for theoretical curve $\gamma = 0.006$. (○) experimental; (—) theoretical. Figure from Sinclair and Topiwala (1970).

A decrease in the temperature of a slow-growing continuous culture increases the steady state viability. At a dilution rate of 0.004^{-1} the steady state viability of *A. aerogenes* increased from 33 per cent to 60 per cent, corresponding to a temperature decrease from 37°C to 25°C (unpublished data of Tempest and Phipps, cited by Postgate, 1973). From an ecological point of view it would be rewarding to extend studies of temperature effects on viability.

Equation 4 suggests that, when a steady state is reached, μ equals D. As pointed out by Sinclair and Topiwala (1970; cf. Powell, 1967), this is not exactly true. In fact, the overall increase in viable cell mass is given by:

$$\frac{dx}{dt} = \mu(s)x - \mu_e x - \gamma x \qquad (11)$$

where μ_e (Powell, 1967) is the amount of cell material not formed per unit time per unit organisms concentration due to a requirement for maintenance energy, and γ is specific death rate. Thus theoretically at all steady states we have $\mu > D$. However, as at the higher growth rates μ_e and γ are generally relatively small, these factors are only important at low growth rates, the actual rate depending on organism and environmental conditions.

Continuous culture studies with partly viable populations of *A. aerogenes* have not provided evidence for emergence of starvation-resistant mutants (cf. Postgate and Hunter, 1962; Postgate, 1973).

2.2 *Competition For A Single Growth-limiting Substrate*

The exploration of the microbial world during the first decades of this century was greatly facilitated after introduction of the enrichment culture technique. However, as this technique was applied in closed culture systems, some questions remained as to the ecological relevance of the information thus obtained. In order to obtain a sizable population of the dominant organism, the concentrations of nutrients applied had to be much higher than those occurring in most natural environments. Therefore selection in these cultures was governed by one growth parameter, the maximum specific growth rate. However, it seemed possible that in nutritionally-poor environments chemoorganotrophic organisms occur which show a high substrate affinity (K_s) for various nutrients, but which show a low μ_{max}. Organisms of this type would never come to the fore in batch culture. An answer to the question whether there is in natural environments a selection towards substrate concentration can be given by competition experiments carried out at different substrate concentrations in the chemostat. As these concentrations are determined by dilution rate (cf. eq. 8, Fig. 3) the type of experiment to be done is simple. Two chemostats are connected to the same reservoir. A high dilution rate (e.g. $0.3h^{-1}$) is adjusted in one vessel, and a low one (e.g. $0.03h^{-1}$) in the other. Both vessels are inoculated with a sample from a ditch, sea water, or a soil suspension. Several experiments of this type have been carried out (Jannasch, 1967; Veldkamp and Jannasch, 1972; Veldkamp and Kuenen, 1973; Jannasch and Mateles, 1974). The results can be summarized as follows. In most cases, after about five volume changes different organisms have become dominant in the two chemostats. This holds for aerobic chemoorganotrophic organisms, both for organic and inorganic growth-limiting substrates. A similar result was obtained in competition for iron among the chemolithotrophs *Thiomicrospira* and *Thiobacillus* (Kuenen *et al.*, 1976; Veldkamp, 1976).

An interesting point is that chemoorganotrophic bacteria isolated at different concentrations of a growth-limiting organic substrate often show similar growth requirements. It is difficult to find an organic substrate which can be attacked by one organism and not by the other. A common feature of bacteria growing relatively fast at extremely low concentrations of limiting substrates seems to be a high surface to volume ratio. These organisms are often spiral-shaped (Veldkamp, 1976).

Preliminary evidence thus indicates that part of the diversity among prokaryotes can be explained in terms of selective advantages related to concentration of growth-limiting nutrients. The results obtained also show that the organisms present in our culture collections, isolated and maintained at high substrate concentrations, are not those responsible for mineralization processes in nutritionally poor environments.

Experiments as described above are easy to carry out and

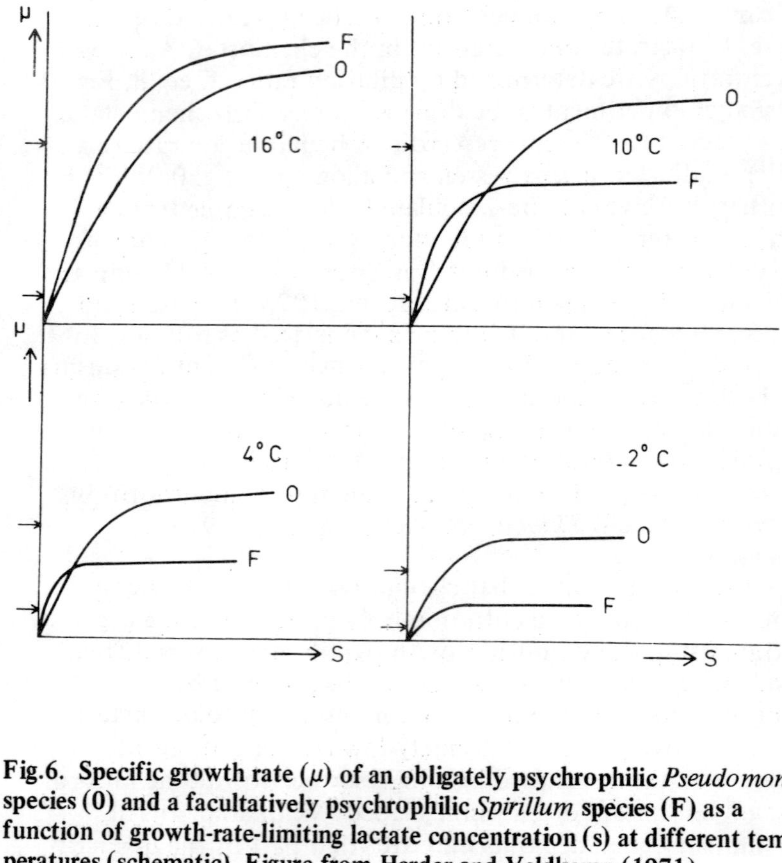

Fig.6. Specific growth rate (μ) of an obligately psychrophilic *Pseudomonas* species (0) and a facultatively psychrophilic *Spirillum* species (F) as a function of growth-rate-limiting lactate concentration (s) at different temperatures (schematic). Figure from Harder and Veldkamp (1971).

therefore suitable for microbiology courses. In this respect, one practical point should be emphasized. Before inoculating the chemostats, the sample should be filtered through a 2μ membrane filter to remove protozoa. Otherwise unintended predator-prey processes almost certainly will occur.

Very few experiments have been done in which, in addition to substrate concentration, a second variable was introduced. An example is shown in Fig.6 (Harder and Veldkamp, 1971). This represents competition between two marine chemoorganotrophic organisms for lactate at different temperatures. The curves are schematic and based on determination of maximum growth rate in batch culture and competition experiments in the chemostat at various dilution rates, indicated by arrows. The organisms used the same range of organic substrates but showed differences in optimum growth temperature, substrate affinity and maximal growth rate. The results indicate that in competition among chemoorganotrophs in permanently cold natural environments (ocean waters, arctic, antarctic) obligately psychrophilic bacteria grow faster than facultative psychrophiles at high as well as very low concentrations of growth-limiting substrates.

In those environments that show seasonal temperature changes it would be expected that there would be a continuous selection towards temperature throughout the year. This phenomenon was encountered by Sieburth (1967) in the Narragansett Bay, off the coast of Rhode Island, in which the water temperature varies from -2°C (end of January) to +23°C (August). During the summer the majority of the bacteria in this area consisted of mesophiles, whereas a high percentage of the winter population was obligately psychrophilic. Facultative psychrophiles were present throughout the year, reaching maximum numbers in spring and autumn when the water temperature was moderate. It seems likely that in competition for a particular growth-limiting substrate between bacteria which show some overlap in growth temperature range, selection towards concentration of a growth-limiting substrate is rather common. A small temperature change may well determine whether a particular microbe will be successful or not (Figure 6).

Another example of application of two variables in competition experiments was given by van Gemerden (1974), who introduced light as the second variable in competition among

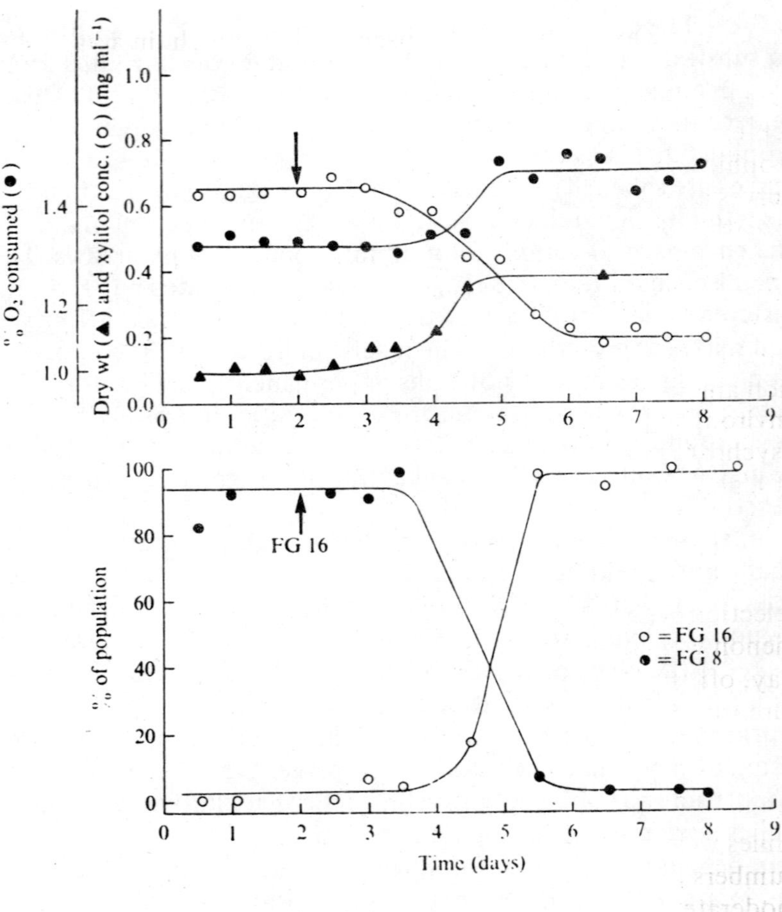

Fig.7. Take-over of *Aerobacter aerogenes* in a chemostat by an evolvant. At the point indicated by the arrow approximately twenty cells of strain FG 16 (a guanine auxotroph of a mutant with improved xylitol dehydrogenase) were added to a steady-state culture containing approximately 10^{12} cells of strain FG 8 (an arginine auxotroph of the ancestral strain) grown on xylitol. The composition of the population was determined by plating on appropriate media. (Hartley *et al.*, 1972).

Chromatium species for sulfide.

Little is known about the biochemical background of differences in μ-s relations in organisms isolated from their natural environment through chemostat enrichments. However, continuous culture studies on evolution of enzymes have provided some elegant examples of the biochemical basis of competitive processes. Hartley *et al.* (1972; cf. Hartley, 1974) evolved the enzyme ribitol dehydrogenase of *Aerobacter aerogenes* to a new specificity for xylitol. A mutant was obtained which was constitutive for ribitol dehydrogenase and which could grow on xylitol by utilizing a side specificity of the enzyme to produce xylulose. However, the K_m (app) for xylitol was very high (1 M) as compared to the K_m for ribitol (1 mM). It was shown that in a xylitol-limited chemostat, the activity of this enzyme was rate-limiting for growth. Thus, when a faster-growing mutant arises in such a culture, one knows where to look for its selective advantage.

The selective power of the chemostat is nicely illustrated by the following experiment (Hartley *et al.*, 1972; cf. Hartley, 1974). A strain of *A. aerogenes*, constitutive for ribitol dehydrogenase, was grown in a xylitol-limited chemostat. After a steady state was established, twenty cells of a mutant strain were inoculated into the culture, which contained approximately 10^{12} cells. The mutant had a two-fold improvement in V_{max}, and a three-fold improvement in K_m (app) for xylitol. Fig.7 shows that within a week the original strain was selectively eliminated. After inoculation, the invader grew faster at the prevailing xylitol concentration, and when at constant dilution rate the xylitol concentration decreased, the original strain became faced with the hopeless situation in which $D > \mu$. The invader strain reached a steady state characterized by a higher population density and a lower xylitol concentration in the culture.

2.3 *Predator-prey Relations*

In natural waters and many other environments, steady states do not occur for any prolonged period of time as chemical com-

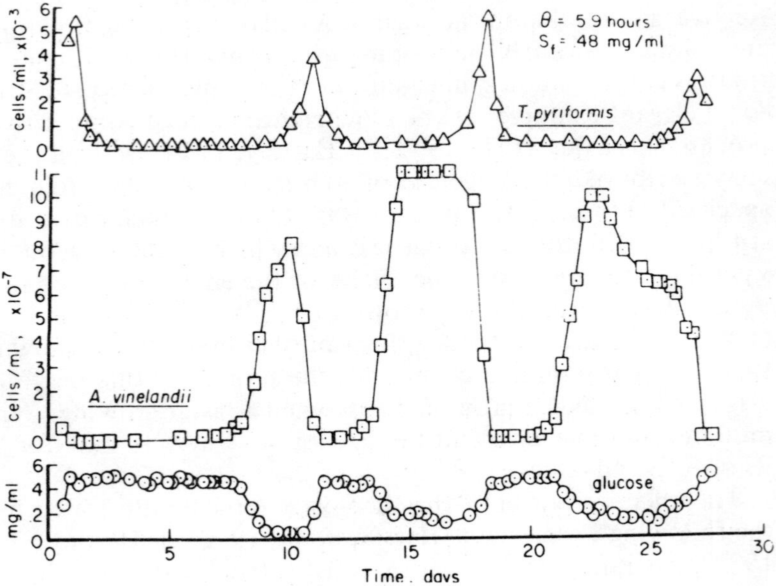

Fig.8. Sustained oscillations in the glucose-*Azotobacter-Tetrahymena* food chain. Holding time 5.9 h corresponds to dilution rate 0.17 h^{-1}. Glucose concentration in feed (s_f) was 48 mg/ml. (Jost *et al.*, 1973a).

position and physical factors such as temperature are continually changing. Fig. 6 showed how changes in concentration of a growth-limiting substrate and temperature changes may affect growth-rates and therefore cause shifts in the dynamic equilibrium among bacterial populations. Another important factor in the regulation of bacterial numbers is predation, as illustrated in the following example.

Jost *et al.*(1973a) grew *Azotobacter vinelandii* in a chemostat under glucose-limitation, and then inoculated *Escherichia coli* into this culture. It turned out that, independent of dilution rate (that is concentration of growth-limiting glucose; cf. eq. 8), *Azotobacter* was selectively eliminated. This was not surprising, as the growth parameters for that growth medium at 25°C were as follows. *E. coli*: $\mu_{max} = 0.32h^{-1}$, $K_S = 1 \times 10^{-7}$mg/ml, and: *A. vinelandii*: $\mu_{max} = 0.23h^{-1}$, $K_S = 1.2 \times 10^{-2}$mg/ml. Even though these organisms have different ecological niches this means that *A. vinelandii* when exposed to glucose-limitation in mixed bacterial culture is apt to be eliminated at low glucose concentrations because of its low substrate affinity. However, as will be shown below, the presence of a predator may favour the *Azotobacter* population.

Jost *et al.* (1973a) studied in a single stage chemostat the relation between pure cultures of glucose-limited populations of *E. coli* and *A. vinelandii* and the predatory ciliate *Tetrahymena pyriformis*. The results for the *Azotobacter–Tetrahymena* mixed population are shown in Fig.8. The behaviour of *E. coli* showed a similar pattern of oscillations when grown at the same dilution rate. Subsequently, Jost *et al.* (1973a) studied population changes when all three organisms were introduced in a glucose-limited chemostat. It turned out that, under these conditions, both bacterial cultures could coexist. In other words, despite the glucose limitation, *Azotobacter* was not eliminated. These experiments illustrate the care that must be taken in interpreting competition experiments, or any other type of experiments in which a single factor affects population levels, in terms of behaviour in a complex natural environment. They also show how continuous cultures can be used in developing a field of research which is still mainly unexplored. Concomitant with its exploration should be the development of mathematical models for such complex systems. Valuable contributions were given by Canale (1969, 1970), Curds (1971a) and Jost *et al.* (1973a). The classical theoretical basis of predator-prey relations was given by Lotka (1925) and Volterra (1926). They developed a model in which oscillations are inherent to the system. However, Tsuchiya *et al.* (1972) and Jost *et al.* (1973a, b) have indicated that oscillations in a single-stage chemostat are not inherent to the predator-prey relation as such. They depend on environmental factors (dilution rate and concentration of the growth-limiting substrate in the feed). With some combinations of these factors, no oscillations occur (cf. Jost *et al.*, 1973b).

Steady states of predator and prey can easily be obtained in a two-stage system in which the bacterial prey grown in the first stage is fed into the second vessel containing the predator. A system of this type was studied by Curds and Cockburn (1971). In the first stage *Aerobacter aerogenes* was grown in a mineral salts-sucrose medium with sucrose as the growth-limiting substrate. Part of this culture was fed into a second vessel where the ciliate *Tetrahymena pyriformis* preyed on *Aerobacter*. It was possible to grow the ciliate at a rate of $0.38h^{-1}$, and the maximum specific growth rate was estimated to be $0.43h^{-1}$. This value was almost twice that obtained in batch culture studies (Curds and Cockburn, 1968). A saturation constant K_S of 11.7mg/l was found for *Tetrahymena*, and a yield coefficient of 0.54 was observed. The data did not allow a definite conclusion as to a dependence of the yield coefficient on growth rate. The yield coefficient of 0.54 of the ciliate was even larger than that of *A. aerogenes*, grown in the first stage (0.45; Curds, 1971a). Thus, about 25% of sucrose-carbon fed into the first stage ended up in ciliate cell material. The ciliate cell volume increased with increasing growth rate, and a linear relation was observed between ciliate feeding rate (μg bacteria/1000 ciliates/h) and the product of mean cell volume and growth rate.

Curds *et al.* (1968) demonstrated that the major role of protozoa in the activated sludge process is to remove the dispersed non-flocculated bacteria, which only appeared in large numbers when protozoa were absent. Subsequent evidence from batch culture studies (Curds and Cockburn, 1968) showed that the predatory activities of ciliated protozoa alone could account for the removal of suspended effluent bacteria. The continuous culture studies of Curds and Cockburn (1971) demonstrated, that under these growth conditions, the feeding and growth rates of *Tetrahymena* were even higher than those observed in batch culture. The usefulness of continuous culture studies on predator-prey relations was further demonstrated by Curds, who used the growth parameters experimentally obtained in continuous culture in computer simulation studies of microbial population dynamics in the activated sludge process (Curds, 1971b), and on predator-prey relationships in a single stage chemostat (Curds, 1971a). A review on the microbial ecology of the activated sludge process was given by Pike and Curds (1971).

3

TRANSFORMATION OF ENERGY AS A FUNCTION OF ENVIRONMENTAL CONDITIONS

Several topics discussed in this chapter are dealt with in detail by Stouthamer (1976) in this series of monographs. Emphasis will be given here on the use that can be made of continuous cultures to study the effect of environmental conditions on generation and utilization of adenosine triphosphate (ATP) in chemoorganotrophic microorganisms.

3.1 *Generation of ATP*

During respiration of chemoorganotrophs, organic compounds are oxidized and part of the energy yielded is conserved in the form of ATP. Most of this is generated in the respiratory chain, in which iron plays a central role. The question thus arises whether in microorganisms exposed to iron limitation the ATP generation may become affected. In a series of elegant experiments with the yeast *Torulopsis utilis* of Garland and coworkers this was one of the questions studied (Light, 1972). When this yeast is grown with excess of nutrients, three sites of energy conservation are present in the mitochondrial respiratory chain. In studying the effect of iron-limitation, *T. utilis* was grown in a chemostat in a medium consisting of minerals with glycerol as C- and energy source. With all other components present in excess, the culture was iron-limited when the ratio of iron to glycerol in the inflowing medium was below 4.8μg atom iron/mol glycerol. Under these conditions no energy conservation was coupled to the almost unaltered electron flow between

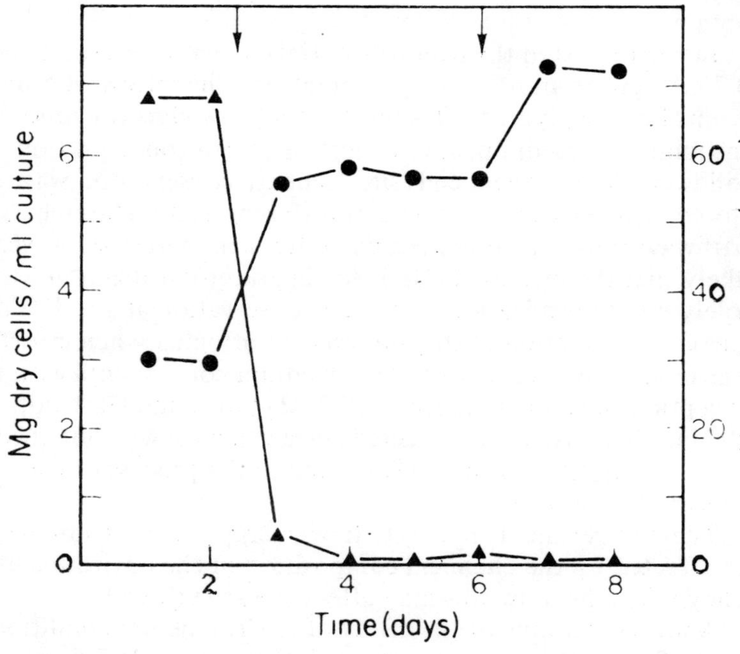

Fig.9. Stepwise increase of $FeCl_3$ in feed of iron-limited *Torulopsis utilis*. ● — ●, culture cell density; ▲ — ▲, glycerol concentration in culture. Input glycerol concentration was 130 mM throughout the experiment. Input iron concentration (increased at the times indicated by the arrows) was 0.3μM on days 0 to 3, 0.64μM on days 3 to 6 and 0.75μM on days 7 and 8. Dilution rate 0.25 h^{-1} (Light, 1972).

intramitochondrial NADH and cytochrome b (site 1). When the $FeCl_3$ concentration in the inflowing medium was stepwise increased a picture was obtained as shown in Fig.9.

Following an increase from 0.3μM to 0.64μM, the population density increased as was expected (cf. eq. 7). A concurrent decrease of the glycerol concentration to a very low value was obtained.

In the next step the iron concentration was increased to 0.75μM, giving an iron to glycerol ratio in the inflow of 5.8μg-atoms Fe/mol glycerol. Despite the very low glycerol concentration in the medium, a 20% increase in cell concentration followed. And in these cells site 1 energy conservation was restored. There was no evidence of differences in metabolic pathways in the different steady states, and it thus seems very likely that the increased efficiency in glycerol utilization was solely due to restoration of energy conservation at site 1. The energy conservation at this site was not affected when growth was limited by a variety of other medium components with the exception of sulphate (Light, 1972; Haddock and Garland, 1971). This was not unexpected since evidence was obtained that an iron-sulfur protein is involved in the process of energy conservation at site 1.

These experiments illustrate how energy conservation may be affected by the chemical composition of the environment. They also show how this may affect the growth yield.

Another example of the effect of environmental conditions on ATP generation was found by de Vries *et al.* (1970) in *Lactobacillus casei*. When grown in a glucose-limited chemostat, the fermentation pattern of this organism is a function of growth rate. At low growth rates 3 moles of ATP are formed per mole of glucose fermented, and at high growth rates only 2 moles are formed. As expected, higher molar growth yields for glucose were found in slow growing organisms.

3.2 *Utilization of ATP*

In a theoretical study on the amount of ATP required for the synthesis of microbial cell material, Stouthamer (1973) showed that the maximal amount of cell material (g dry weight) that can be formed per mole ATP (Y_{ATP}^{max}) is dependent on the growth medium. For a chemoorganotrophic microbe utilizing glucose as C- and energy source and provided with various supplements, Y_{ATP}^{max} varies according to these calculations from 28.6 to 32.1.

The ATP formed in dissimilatory processes is not only used for biosynthetic purposes, but also for a number of other processes. The ATP needed for transport and for turnover of messenger RNA is included in the above figure. All other energy required for processes not directly related to growth is collectively considered as "maintenance energy".

Part of the energy conserved in ATP during dissimilation may not be used at all for either biosynthesis or maintenance. The efficiency of ATP utilization for growth depends not only on the needs for maintenance purposes, but also on the degree of coupling between dissimilation and assimilation. The continuous culture technique is very useful in studies on maintenance energy requirement and the coupling of catabolism to anabolism, as will be shown in the following paragraphs.

Growth Under Energy-limitation

The maintenance requirement can be observed in a chemoorganotrophic organism when it is grown in a chemostat under energy-limitation. At low dilution rates the percentage of available ATP needed for maintenance purposes becomes relatively high, and under these conditions the cell concentration becomes less than predicted by theory as shown in Fig.3. The yield coefficient is not, in fact, a constant, but depends on growth rate (Fig.13).

A mathematical expression for the maintenance requirement was given by Pirt (1965; cf. van Uden, 1969; Pirt, 1975), who derived the following equation:

$$\frac{1}{Y} = \frac{1}{\mu} \cdot m_S + \frac{1}{Y_G} \qquad (12)$$

where the observed molar growth yield Y = g dry weight/mole of substrate used; the maintenance coefficient m_S = moles substrate/g dry weight/h; μ is specific growth rate in h^{-1}, and Y_G is the molar growth yield corrected for the maintenance requirement.

Pirt (1965) showed that the equation is in good agreement with experimental data. When $\frac{1}{Y}$ is plotted against $\frac{1}{\mu}$ a straight line is generally obtained, at least when extremely high and low growth rates are left out of consideration. Its slope equals the maintenance requirement, and this requirement thus appears to be independent of growth rate in most cases. The intercept with the ordinate is $\frac{1}{Y_G}$, and Y_G thus can be calculated. Its value is of course not necessarily equal to the theoretical amount of cell material that can be formed from the ATP produced in dissimilation after a correction for the maintenance requirement has been made. In fact, it is generally much smaller.

The maintenance requirement is highly dependent on environmental conditions. An example is given in Fig.10, which shows the behaviour of the yeast *Saccharomyces cerevisiae* in a glucose-limited chemostat as affected by the NaCl concentration in the growth medium.

A similar increase in maintenance requirement may well occur in bacteria flowing with river water into an estuary where they are exposed to rising NaCl concentrations. This increased maintenance requirement may become fatal when the amount of energy source available per unit time is lower than needed for covering the increased maintenance requirement.

Equation 12 cannot be used in cases where ATP generation is a function of growth rate, as was shown above for glucose-limited *L. casei*. In that case the following equation (de Vries *et al.*, 1970) should be used:

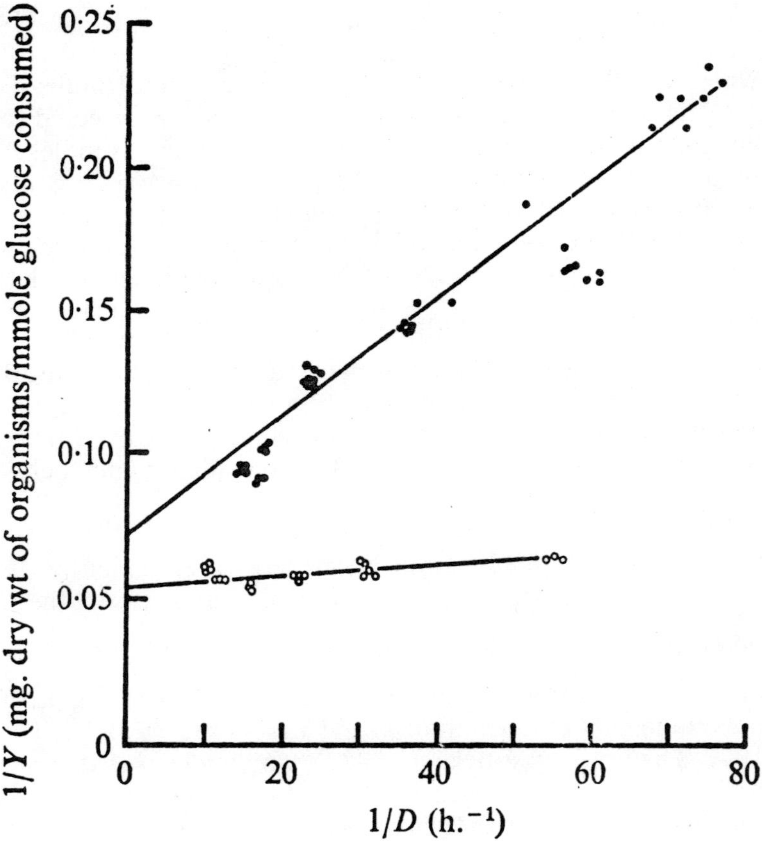

Fig. 10. Reciprocal plots relating yield (based on glucose consumed) to dilution rate in carbon-limited continuous cultures of a respiratory-deficient mutant of *Saccharomyces cerevisiae* IGC 3507. ●, Data for culture with 1.0 M-NaCl; ○, data for culture without NaCl (Watson, 1970).

$$\frac{1}{Y_{ATP}} = \frac{1}{\mu} \cdot m_e + \frac{1}{Y'_{ATP}} \tag{13}$$

where the observed Y_{ATP} is g dry weight produced/mole of ATP used; Y'_{ATP} is the growth yield/mole ATP, corrected for energy required for maintenance purposes and m_e is moles ATP required for maintenance/g dry weight/h. In the experiments of de Vries *et al.* (1970), the Y_{ATP} found for *L. casei* varied from 18.7 at low dilution rates to 23.5 at high dilution rates; m_e was 1.5×10^{-3}, and a Y_{ATP} value of 24.3 was calculated. The latter value is quite high as compared to the theoretical Y^{max}_{ATP} of around 28. However, there are many batch culture data available which show rather low Y_{ATP} values. (Payne, 1970; Forrest and Walker, 1971). Under these conditions cells grow at maximal rate with excess substrate, so the large discrepancy between observed Y_{ATP} and theoretical Y^{max}_{ATP} cannot be accounted for by the relatively small maintenance energy requirement.

It seems appropriate to assume that the gap between the observed Y_{ATP} and the theoretical Y^{max}_{ATP} is not only due to use of ATP for maintenance purposes, but also to some degree of uncoupling between catabolism and anabolism (cf. Belaich *et al.*, 1972; Lazdunski and Belaich, 1972), which means that some ATP is not used at all, either for growth or maintenance. This can be accounted for as follows (Harder, 1974):

$$Y'_{ATP} = K \cdot Y^{max}_{ATP} \tag{14}$$

where K is the coefficient that gives the degree of coupling between anabolism and catabolism. When K = 1, the coupling is maximal and the for maintenance corrected Y_{ATP} then equals the theoretically calculated Y^{max}_{ATP}.

Equations 13 and 14 then thus give:

$$\frac{1}{Y_{ATP}} = \frac{1}{\mu} \cdot m_e + \frac{1}{K.Y_{ATP}^{max}} \qquad (15)$$

Equation (15) was used by Harder and van Dijken (1976) to calculate cell yields of organisms growing on methane. The values of m_e and K vary with environmental conditions. However, they are independent of growth rate; apparently these two parameters are not affected by changes in the concentration of the growth-limiting substrate.

The maintenance requirement can be determined accurately only in continuous culture. Practically all studies of this type were carried out with chemoorganotrophic organisms limited by the C- and energy source. However, equations 12 and 16 (section 3.3) can also be applied with respect to the C- and energy source for cultures limited by another nutrient (Stouthamer and Bettenhaussen, 1973). With both types of nutrient limitation, the amount of cell material produced per mole of C- and energy source consumed is a function of growth rate. However, available data (Stouthamer, 1976) show that the molar growth yield is higher in C-limited cultures, except at growth rates near maximal. The lower the growth rate, the larger the difference in molar growth yield between C-limited cultures and cultures limited by another nutrient. This means that in natural environments where low growth rates are the rule, the efficiency of biomass production in mineralization processes is considerably higher in cells of chemoorganotrophic bacteria limited by C- and energy sources as compared to cells limited by other nutrients. Thus, whatever the growth-rate-limiting substrate, the activities of chemoorganotrophic bacteria shift from biomass formation to nutrient cycling with decreasing growth rates. And the effect is least pronounced in C-limited populations.

Observations of a similar kind were obtained with *Hydrogenomonas* when grown autotrophically under H_2, O_2 and CO_2 limitation, respectively (Bongers, 1970). The energy of hydrogen oxidation was most efficiently used under H_2 and O_2 limitation. In CO_2-limited cultures the cells showed excess oxidative activity which increased with decreasing growth rates. In the complete absence of CO_2 no less than 40% of the maximal oxidative activity remained.

Even though C- and energy limitation in chemoorganotrophs promotes an efficient use of this source, the actual cell yield may be highly dependent on other environmental conditions. An example was given by Harrison and Loveless (1971; cf. Harrison, 1973) who grew *Aerobacter aerogenes* in a glucose-limited chemostat. The growth yield coefficient was unaffected by temperature between 15° and 36°C, and no significant effect of pH was observed between 6.6 and 8.2. However, when the oxygen supply was diminished and the culture, growing at μ = $0.21 h^{-1}$, was only just oxygen-limited the cell yield coefficient for glucose decreased by nearby 40%. A concomitant increase in Q_{O_2} (mmole 0_2/g cells per h) of the culture was observed, whereas the dissolved carbon concentration in the culture was about the same as the fully aerobic value.

Growth At Submaximal Rate With Excess Of Carbon- And Energy Source

Several continuous culture studies have been made on coupling between catabolism and anabolism in chemoorganotrophs growing under nutrient limitation other than the C- and energy source. When the specific rate of carbon consumption is plotted against the specific growth rate, a straight line is obtained (e.g. Baschnagel-DePamphilis and Hanson, 1969; Jensen and Neidhardt, 1969; Stouthamer and Bettenhaussen, 1973). A typical example is presented in Fig.11, which shows aerobic growth of *A. aerogenes* in the presence of excess of the C- and energy source histidine.

Histidine catabolism appeared to be only partially coupled to growth rate. The coupling was achieved by catabolite repression of histidase and also by a slightly decreased function of this enzyme at low growth rates *in vivo*. Nevertheless, at all submax-

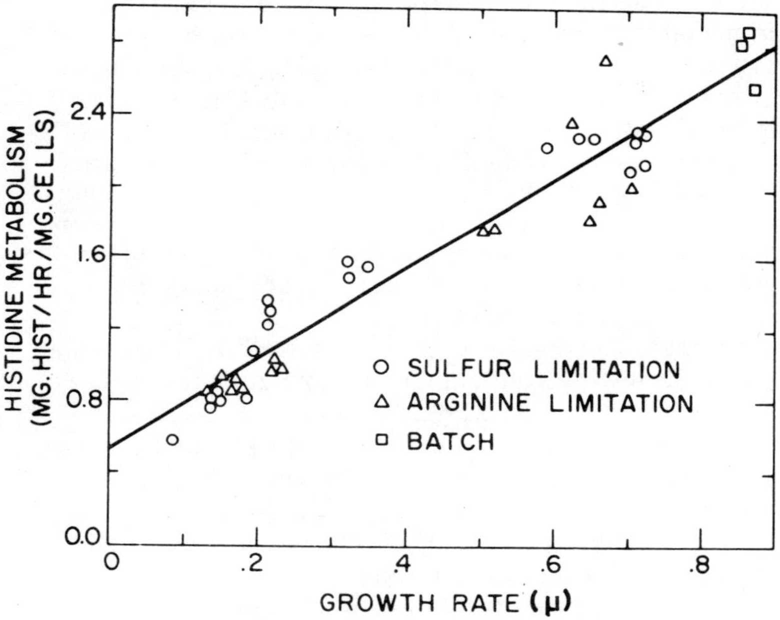

Fig.11. Metabolism of L-histidine as a function of growth rate. The specific rate of L-histidine metabolism (expressed as milligrams of L-histidine degraded per hour per milligram, dry weight, of cells) is plotted as a function of the specific growth rate μ (h^{-1}), for many cultures of *A. aerogenes* LC1 in steady-state growth at 37°C in a chemostat with sulfur limitation (○) or with L-arginine limitation (△), or in batch culture (□). The solid line was drawn by the method of least squares as a line of regression of y on x (Jensen and Neidhardt, 1969).

imal growth rates there was overproduction of this enzyme. Experimental evidence showed that the excess consumption of histidine was not merely a matter of covering the maintenance energy requirement, but rather an overall imbalance between substrate utilization and biosynthetic demands. The failure of catabolite repression to coordinate synthesis of histidase strictly to growth rate could probably be compensated for by the ability of the cells to convert potentially repressing catabolites to non-repressing end products which were released in the culture medium. In fact, the ratio in which these products were released was a function of growth rate.

One example showing no coupling of dissimilation to assimilation at all seems an exception rather than a rule. It was found by Rosenberger and Elsden (1960) in *Streptococcus faecalis* when grown in a rich medium under tryptophan-limitation with excess glucose. At all growth rates the specific glucose consumption rate was maximal. But in this case glucose served as energy source only, as practically all cell carbon was derived from monomers in the growth medium.

Growth under substrate limitation other than the C- and energy source often leads to synthesis of reserve material such as polyglucose or poly-β-hydroxybutyrate (PHB). In fact, one of the deviations from continuous culture theory as given in Fig.3, in which abnormally high cell concentrations (dry weight/ml) at low growth rates were found with N-limited populations growing with excess glucose (Herbert, 1961), occurred because the cells were stuffed with reserve material. It should be emphasized that whether or not reserve material is synthesized in the presence of excess C- and energy source depends on the kind of growth limitation. For instance, *Aerobacter aerogenes* forms polyglucose under NH_4^+-, PO_4^{3-}– and SO_4^{2-}–limitation, but hardly any under K^+- and Mg^{2+}-limitation (Tempest *et al.*, 1965; Dicks and Tempest, 1966).

Continuous culture with excess of the carbon-energy source has revealed several regulatory processes. Particularly interesting is the formation of PHB in *Azotobacter beijerinckii* under various

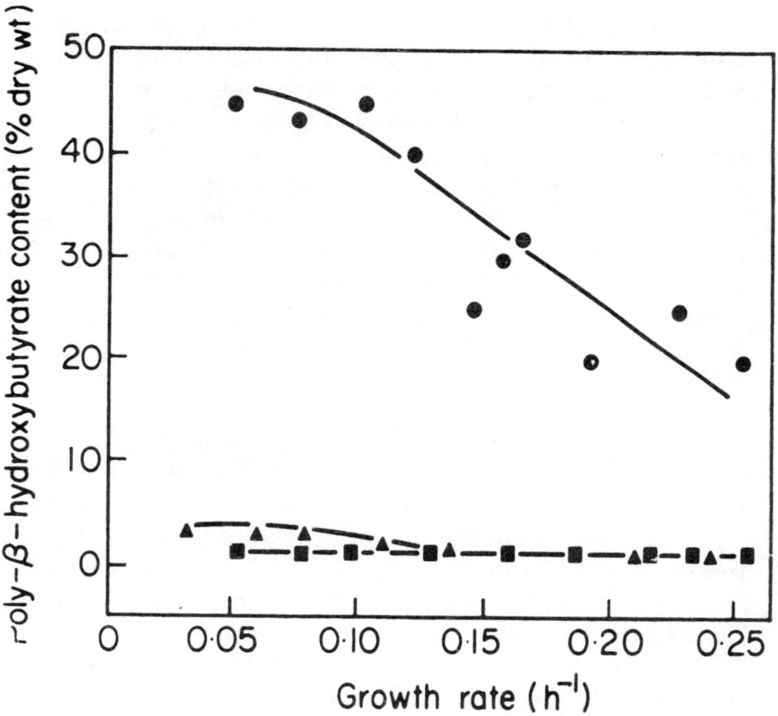

Fig.12. Effect of growth-limiting substrate and dilution rate on the poly-ß-hydroxybutyrate content of *Azotobacter beijerinckii*. Limitations were: oxygen, ●; nitrogen, ■; carbon (glucose), ▲. From Senior *et al.* (1972).

nutrient limitations, as shown in Fig.12.

In the presence of excess glucose, large quantities of PHB are formed in O_2-limited cultures, but not when growth is limited by N_2. Under O_2-limitation PHB accumulation functions probably as an "electron-sink" for excess reducing power, as under these conditions the cells are not able to oxidize NAD(P)H produced in glucose dissimilation as fast as in N_2- or C-limited cultures with excess of oxygen. Further details of the regulation of PHB metabolism are given by Dawes and Senior (1973).

The behaviour of *Azotobacter chroococcum* grown under various growth limitations was studied by Dalton and Postgate (1968, 1969) and reviewed by Hill *et al.* (1972). When grown in a medium with excess of minerals, mannitol and oxygen, and N_2 as the N-source, the culture was N_2-limited at all growth rates. And it was shown that this was due to some intrinsic property, as variations in pN_2 did not affect cell concentration as would be expected according to equation 7. At high dilution rates a selection of fast-growing variants of the population was observed.

Dalton and Postgate (1968, 1969) proposed that the regulation of the N_2-fixation rate is part of a mechanism for protecting the nitrogenase system from damage by oxygen. The enzyme complex consists of two proteins, one of which is highly sensitive to oxygen (Postgate, 1972). It was suggested that N_2-fixing *A. chroococcum* has two protective mechanisms for operating the oxygen-sensitive enzyme system in the presence of air. Firstly, respiratory protection and secondly, conformational protection. One of the observations that support the first hypothesis was the extreme sensitivity to oxygen of C-limited cells. When grown under these conditions an exceptionally high maintenance coefficient was found (1.06 g mannitol/g dry wt organisms/h) as compared to ammonia-grown C-limited cells ($m = 0.04$). Therefore with decreasing growth rates, C-limited cells when fixing nitrogen show a dramatic decrease in molar growth yield. As pointed out by Dalton and Postgate (1969), the term "C-limited" has a rather special meaning in this case, where limitation by one metabolic pathway (C) may lead to intoxication of a second pathway (N) by a substrate for a third (O).

Drozd *et al.* (1972) found that increasing concentrations of ammonium ions in the medium of N_2-fixing sulphate-limited continuous cultures of *A. chroococcum* caused a proportionate repression of nitrogenase activity.

3.3 *Growth Rate And Growth Yield*

The relation between specific growth rate and growth yield can be expressed as follows (Schulze and Lipe, 1964; van Uden, 1967, 1969; Pirt, 1975):

$$q = \frac{\mu}{Y} = \frac{\mu}{Y_G} + m_s \qquad (16)$$

where q is the specific substrate uptake rate (mmoles of substrate used/mg dry weight/h); the observed molar growth yield Y = g dry weight/mole of substrate used; Y_G is the molar growth yield corrected for maintenance requirement; the maintenance coefficient m_s = mmoles substrate/mg dry weight/h; μ is specific growth rate in h^{-1}. In this equation $\frac{\mu}{Y_G}$ represents rate of substrate consumption for growth, and m_s the rate of consumption for maintenance purposes.

Fig.13 shows q and Y as a function of μ as observed in *Pseudomonas oxalaticus*. The cells were grown in a mineral medium with oxalate as the limiting substrate, and Fig.13 shows that extrapolation of q to zero growth rate gives an intercept with the ordinate, which equals the growth-rate-independent maintenance requirement.

As was discussed in Chapter 2, the result of competition for growth-limiting substrates is determined by growth rate. One could imagine two chemoorganotrophic organisms showing exactly the same relationship of μ to s for a particular carbon- and energy source, of which one organism shows a lower molar growth yield than the other. Thus, the yield factor is not necessarily critical for growing cells competing for a substrate that limits growth rate. It may become important, however, when the growth limitation becomes absolute and no growth is possible anymore. The viability of bacterial populations exposed to starvation may decrease quite rapidly (Strange and Hunter, 1967). When two populations enter a period of starvation and

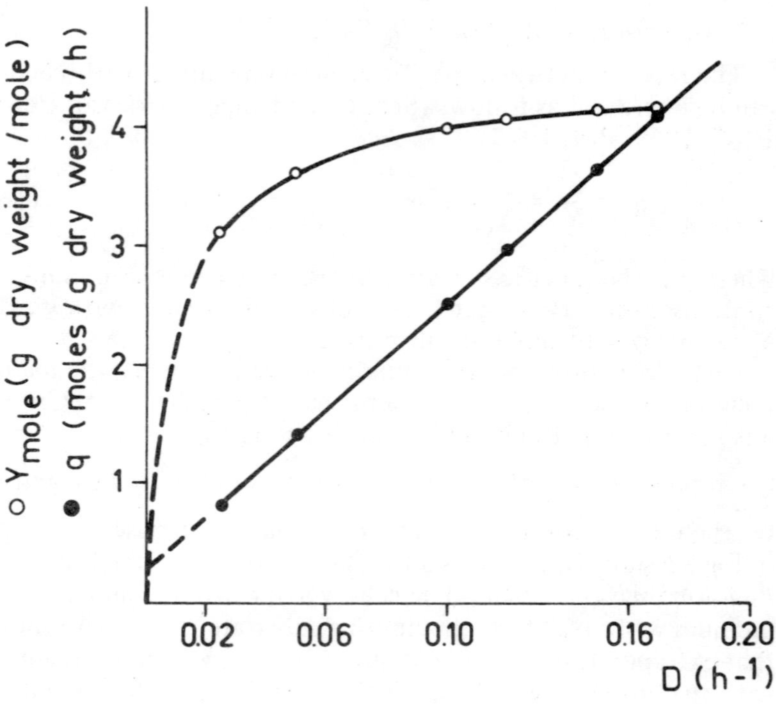

Fig.13. Molar growth yield (Y) and specific oxalate uptake rate (q) as a function of growth rate (μ). Steady-state values obtained at various dilution rates in oxalate-limited *Pseudomonas oxalaticus*. (Harder, 1974).

the death rates are equal, the population having the highest initial population size has the best chances for survival. A high yield is only effective in this respect, when it corresponds to a high number of cells. And this depends on the mean cell mass of the population. A high yield does not necessarily correspond to a high population density, a fact which is illustrated by pure culture studies in which the mean cell mass of a bacterial culture decreases with growth rate. Data given by Herbert (1961) for a glycerol-limited *Aerobacter aerogenes* show that the growth yield (g cells formed/g glycerol consumed) decreases by a factor 2 when the dilution rate is decreased from 0.9 to 0.05 h^{-1}. However, at the same time, the mean cell mass decreases by a factor 3. This means that although at D = 0.05 h^{-1} only half the amount of cell material is formed from glycerol, the number of cells produced per g glycerol consumed is larger at the low dilution rate.

Predation should affect the growth yield of bacterial populations which grow under nutrient limitation. Provided the rate at which the growth rate-limiting substrate becomes available remains constant, introduction of a predator into a chemostat culture gives an increase in the concentration of the growth-limiting substrate. This results in an increase in the specific growth rate (Fig.2), which in turn gives an increase in growth yield (Fig.13).

4

RESPONSES TO A CHANGING ENVIRONMENT

Unicellular organisms are generally exposed to a continually changing milieu, and their metabolism is especially adapted to respond quickly to such changes. The chemostat is a useful tool to study transient stage behaviour, both on the metabolic and population level. A few examples are given below of responses to chemical and physical changes by cells growing at submaximal rate under steady state conditions.

4.1 *Chemical Changes In The Environment*

When the dilution rate of a steady state culture is increased to a value not exceeding the critical value D_c (cf. Fig.3 and eq.3), a new steady state will be reached after some time. Upon a sudden increase in dilution rate, the cells are exposed to an increased concentration of the substrate that limits growth rate. And the question is whether, and to what extent, the biosynthetic machinery can give an immediate response to the release of substrate limitation. This was studied by Mateles *et al.* (1965), who developed a technique for the accurate estimation of unsteady state growth rates. For this purpose the following equation was used:

$$\mu = D - \frac{dN/dt}{(N_o - N)} \qquad (17)$$

where N represents the concentration of the N-source in the growth vessel, and N_o is its concentration in the medium reservoir. When N is continuously measured, then with the aid of equation 17 the instantaneous growth rate can be calculated.

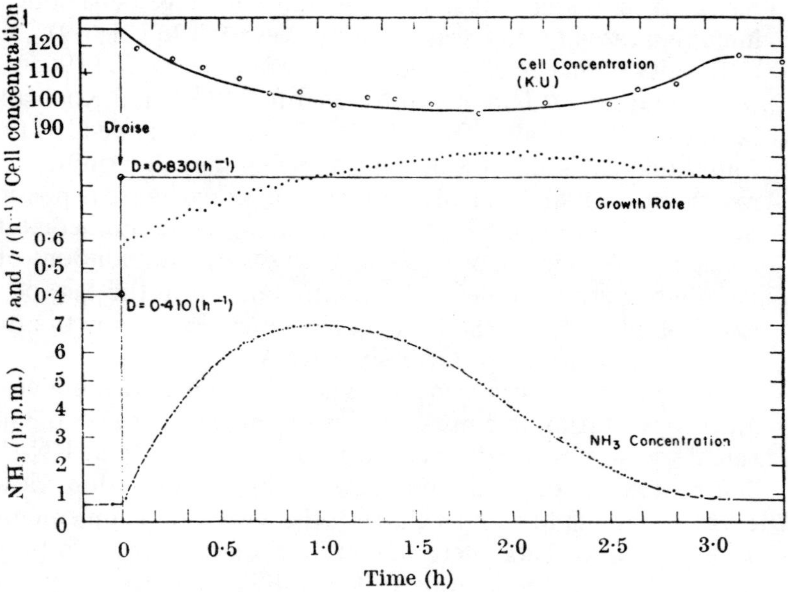

Fig.14. Transient stage behaviour of N-limited *Escherichia coli*, after a sudden increase in dilution rate from 0.4 to 0.8 h^{-1}. (Mateles *et al.*, 1965).

An experiment of this type was carried out with N-limited *Escherichia coli* growing at D = 0.41 h^{-1} in a mineral medium with glycerol. The transient stage behaviour of the culture after a sudden increase of D to 0.83 h^{-1} is shown in Fig.14, which shows that the cells were able to increase their growth rate immediately to 0.6 h^{-1}. This increase was then followed by a prolonged gradual increase. Obviously, N-limited cells growing at a rate of 0.4 h^{-1} had a considerable overcapacity in biosynthetic and transport abilities. Also, the energy generating system could immediately respond to the increased energy requirements. A mathematical analysis of transient stage behaviour, including these results, was given by Powell (1967, 1969).

Results similar to those obtained by Mateles *et al.* (1965) were obtained by Jensen and Neidhardt (1969), and Koch and Deppe (1971; see also Koch, 1971). Koch and Deppe (1971) found that the kind of nutrient limitation may determine whether or not an immediate increase in growth rate is possible. Slow growing ($\mu < 0.1\ h^{-1}$) *Escherichia coli* could immediately change the efficiency of protein synthesis per unit amount of RNA when grown under C-limitation, but not in the case of sulfate-limitation. In the latter case an increase in growth rate exactly paralleled a *net* increase of RNA.

Harvey (1970) observed that when *E. coli* is grown under glucose-limitation in a minerals-glucose medium at $D > 0.3\ h^{-1}$, there was no immediate increase in the specific rate of RNA and protein synthesis upon release of glucose-limitation. However, when, in addition to glucose, the cells were supplemented with 20 amino acids, there was an immediate increase in the rate of RNA synthesis, soon followed by an increase in growth rate. Cells growing at $D < 0.3$ showed a different behaviour, in that release of glucose-limitation resulted in an instantaneous though limited increase in growth rate.

In the minerals-glucose medium the respiratory activity was constant and maximal over the dilution rate range of 0.3 to $0.55\ h^{-1}$. However, the specific activity of glutamic dehydrogenase and glutamic-oxalacetic transaminase at these dilution rates was a linear function of growth rate. And extrapolation gave a line passing through the origin. The fact that cells growing at low dilution rate were able to an instantaneous increase of growth rate upon shift-up was explained in terms of modulation of enzyme activity.

With respect to incorporation of NH_3-N into cell material, the continuous culture studies with *Aerobacter aerogenes* of Tempest and coworkers are of particular interest (Tempest *et al.*, 1970). In the presence of excess NH_3-N, the incorporation occurs in the usual way through reductive amination of

∝-ketoglutarate. However, the K_m of glutamic dehydrogenase for NH_3 is rather high (3 to 4 mM). In fact, much higher than the internal NH_3-concentration (< 1 mM) of cells growing at D = 0.3 h^{-1} in an ammonia-limited chemostat. Further experiments showed that *A. aerogenes* when grown under these conditions applied a different metabolic pathway for glutamate synthesis, of which the first enzyme, glutamine synthetase, showed a K_m < 1 mM for ammonia. The newly discovered pathway appeared to occur also in a variety of other microbes when grown at very low NH_3-concentrations (Meers *et al.*, 1970). The synthesis of glutamine synthetase was found to be regulated by the intracellular ammonia concentration (Meers and Tempest, 1970).

For further information on enzymic activities in nutrient-limited microbes, the reviews given by Clarke and Lilly (1969), and Dean (1972) should be consulted.

One of the conclusions that can be drawn from the experiments mentioned is that microbial cells when growing at submaximal rate under nutrient limitation often can increase their growth rate considerably and instantaneously upon release of the limitation. This is not unexpected when considering the fact that selection during evolution did not act upon cells growing under steady-state conditions. It would be of interest to extend studies on the degree of "preparedness" of microbial cells as a function of growth rate and kind of substrate limitation.

Another aspect of responses to a changing environment is related to the number of C- and energy sources that can be utilized by chemoorganotrophic organisms. Many microbes can utilize a whole range of such compounds. Since the classical batch culture experiments of Monod (1942) it is known that when a mixture of sugars is presented to a bacterial culture, preferential use of one sugar is observed. Only after its consumption, are the inducible enzymes necessary for the utilization of another sugar formed. This phenomenon, termed diauxie by Monod (1942), led to the theory of catabolite repression (Neid-

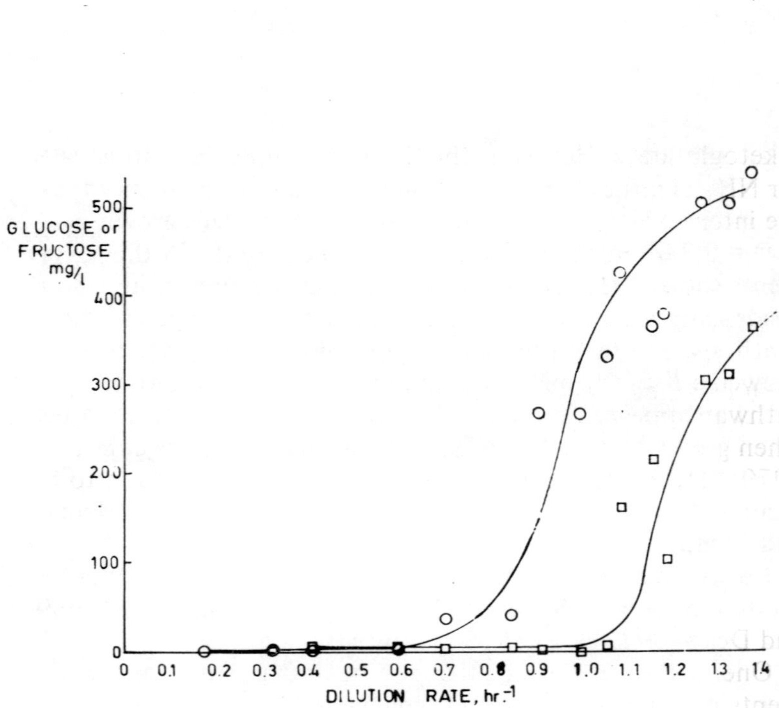

Fig.15. Substrate utilization curves for *Escherichia coli* with mixed feed of the limiting substrates glucose and fructose. Fructose ○; glucose □. (Mateles *et al.*, 1967).

hardt and Magasanik, 1956; Magasanik, 1961) and to many investigations on its molecular basis. Less attention has been paid to microbial behaviour at very low concentrations of mixed C- and energy sources. Such studies are particularly relevant with respect to microorganisms thriving in nutritionally-poor natural environments. Under these conditions more than one C- and energy source might well be utilized simultaneously, and this was confirmed by experiments of Mateles *et al.* (1967) with *Escherichia coli* and *Pseudomonas fluorescens*. The organisms were grown under C-limitation with two carbon sources, glucose and fructose. And it was found that at low dilution rates, that is at low concentrations of the substrates that limited growth-rate, both sugars were used simultaneously. However, at high dilution rates glucose was used preferentially. An example, obtained with *E. coli* is shown in Fig.15. Similar results were obtained with a mixed culture of organisms derived from river water.

The results of Mateles *et al.* (1967) were confirmed by Standing *et al.* (1972) who grew *E. coli* in the presence of glucose and xylose. In batch culture diauxic behaviour was observed, whereas in the chemostat under C-limitation both sugars were used simultaneously.

In a study on the control of mixed substrate utilization in continuous cultures of *E. coli* (Silver and Mateles, 1969), catabolite repression was shown to be the primary mechanism of control of β-galactosidase level and lactose utilization.

In summary, the few continuous culture studies made on utilization of a mixture of C- and energy sources when these are limiting growth rate, show that at very low concentrations there is a release of repression and simultaneous consumption of the mixture. It would be of interest to extend these studies to mixotrophic organisms which can grow as chemoorganotrophs when organic compounds are present and as chemolithotrophs in an inorganic medium.

The literature on mixed substrate utilization has been reviewed by Harder (1976).

4.2 *Physical Changes In The Environment*

Few studies have been made of responses to changes of physical factors by cells growing in continuous culture. Most of these pertain to temperature, and these studies were reviewed by Hunter and Rose (1972).

Growth at the same rate over a range of temperature is possible only in cells growing at submaximal rate. The lower the growth rate, the larger the temperature range at which this can be realized on both sides of the temperature optimum for growth.

TABLE 1

RNA content and ratio ribosomal RNA/total RNA of a *Pseudomonas* spec. grown at a constant rate at different temperatures and at a constant temperature at different rates (Harder and Veldkamp, 1969).

$D = 0.05\ h^{-1}$, temp. = 5 - 19°C			temp. = 14°C, $D = 0.025 - 0.180\ h^{-1}$		
temp. °C	total RNA µg/mg dry weight	ribosomal RNA % of total RNA	D h^{-1}	total RNA µg/mg dry weight	ribosomal RNA % of total RNA
5	205	78	0.025	136	75
8	180	78	0.050	145	78
11	160	78	0.100	166	78
14	145	78	0.130	181	79
16	153	79	0.150	188	79
18	168	81	0.180	202	79
19	154	80			

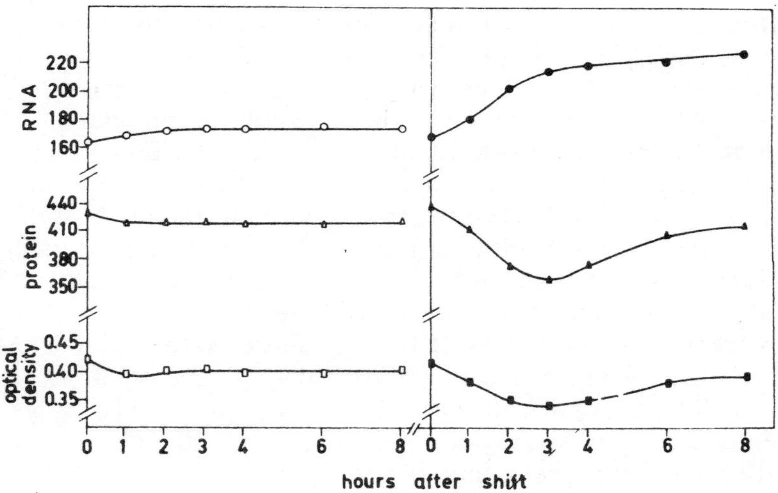

Fig.16. Effect of simultaneously shifting temperature and dilution rate to 14°C and 0.10 h^{-1}, on cultures previously grown at 10°C (open symbols) and 18°C (closed symbols) at a dilution rate of 0.05 h^{-1}. Optical density of the culture was measured at 430 nm; protein is expressed as μg/ml of culture, RNA as μg/mg dry weight of cells. (Harder and Veldkamp, 1969).

Studies on the macromolecular composition of bacterial cells have shown that when these are grown at constant submaximal rate at optimal temperature for growth and the temperature is lowered, the response to decreased reaction rates is to produce more ribosomes. In fact, RNA concentration increases with decreasing temperatures (Tempest and Hunter, 1965; Harder and Veldkamp, 1967). Similarly, when cells are grown at constant temperature and the growth rate increases, an increase in cellular RNA is also found (e.g. Herbert, 1961; Schaechter, 1973). An example of such observations is summarized in Table 1, which shows data obtained with a pseudomonad with optimum and maximum temperature for growth of 14° and 20°, respectively. Cells were grown in a mineral medium with lactate as growth-limiting substrate. The unexpected feature of these experiments was that cells growing at constant submaximal rate showed an RNA increase both below and above optimum temperature for growth. Obviously, these increases were reflections of different phenomena. Apparently RNA synthesized between 14° and 18° was less efficient in protein synthesis than that of cells grown at suboptimal temperatures. Direct evidence for this inefficiency was obtained as follows. Cells growing at 10° and 18° at a dilution rate of 0.05 h^{-1} showed the same content of RNA species as cells growing at 14° at a dilution rate of 0.10 h^{-1}. If the properties of ribosomes synthesized at 10° and 18° at $\mu = 0.05\ h^{-1}$ were similar, then a sudden and simultaneous change of temperature to 14° and dilution rate to 0.1 h^{-1} should not give rise to different behaviour of both types of culture. In fact, it did, as shown in Fig.16.

The 10°-grown cells behaved as expected, but the cells cultivated at 18° were in trouble during the first few hours after the shift. A transient stage was obtained in which $D > \mu$, as reflected by a decreasing number of cells (mg protein/ml, and optical density) and the fortunate cells that stayed in the culture vessel immediately responded by increased RNA-synthesis. It thus seemed that indeed RNA synthesized at 18° was less effective. Several other explanations were ruled out, and the inefficiency was confirmed in enzyme-induction experiments.

The properties of the protein-synthesizing machinery can also be studied in cells grown in batch culture at maximal rate and various temperatures. Very informative studies of this kind have been made (Malcolm, 1968a, b). But for studies of responses of cells growing at submaximal rate to sudden changes in temperature, a chemostat is needed.

The field of research of responses to other physical changes by cells growing at submaximal rate is largely unexplored. As an example, little is known in this respect of the behaviour of photosynthetic microbes to changes in light quality and quantity.

5

MORPHOLOGY AND CELL COMPOSITION

Both morphology and composition of microbial cells are highly dependent on environmental conditions. And as these conditions are well-defined in continuous cultures, this method of cultivation is very useful when studying cell morphology and composition. In batch culture, changes in cell morphology and composition due to nutrient limitation can only be observed in the short period between the end of the exponential phase and the beginning of the stationary phase; but in continuous culture, the properties of cells at growth rates ranging from near zero to near maximal can be studied for any length of time.

Reviews on the chemical composition of microorganisms as affected by the environment were given by Herbert (1961) and Neidhardt (1963). In the discussion below, emphasis is given to morphology and cell composition as a function of growth rate.

5.1 *Morphology As A Function Of Growth-rate*

Fast-growing cells have a greater mean cell mass than slow-growing cells. This was shown by Herbert (1958, 1961) for *Bacillus megaterium* and by Schaechter *et al.* (1958) (see also Maaløe, 1960; Schaechter, 1973) for *Salmonella typhimurium*. These observations have been confirmed with other organisms; an example is given in Fig.17, which shows that not only the surface to volume ratio, but also the appearance of bacterial cells may be a function of growth-rate. *Spirillum* species may be difficult to recognize as such at high growth-rate. Another example

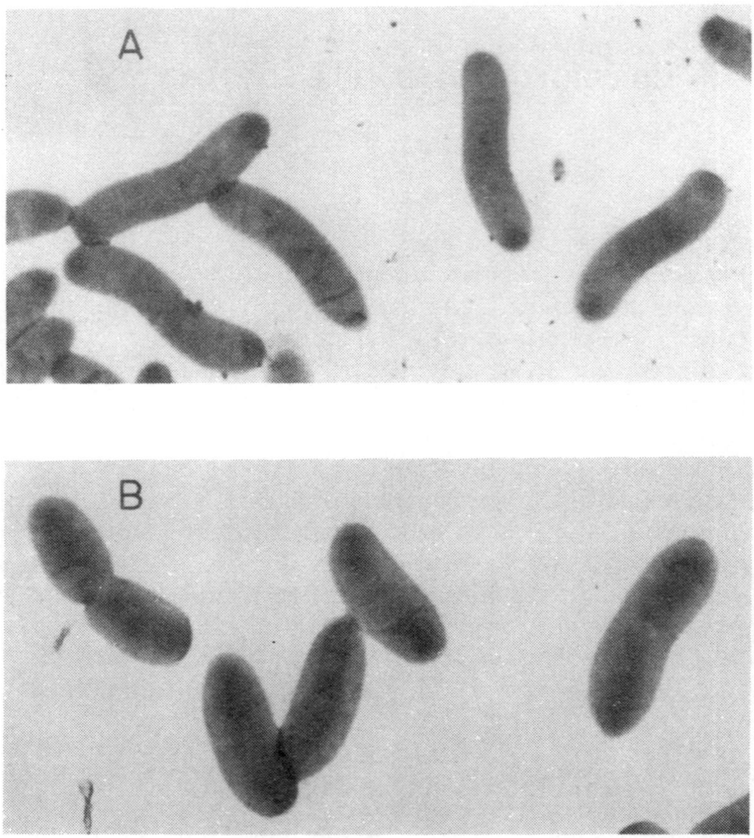

Fig.17. ***Spirillum*** **species grown in chemostat at different rates: A, 0.03 h^{-1}; B, 0.30 h^{-1}. Cells were grown aerobically in a minerals-lactate medium, with lactate as growth-limiting substrate. Maximum specific growth rate in this medium: 0.36 h^{-1}. Stain used: 1% uranyl acetate. Magnification 11.000 x (Photo by courtesy of A. Matin and M. Veenhuis).**

is that of *Arthrobacter* species, the cells of which are coccoid when growing at low rate, whereas fast growing cells are rod-shaped (Luscombe and Gray, 1971).

In *Hyphomicrobium* the stalk becomes longer with decreasing growth rate. The mean stalk length of cells grown at a dilution rate of 0.02 and 0.08 h^{-1}, respectively, was 2.8 and 1.1μ. The corresponding sizes of the mother cells were 0.8 x 0.5μ and 1.0 x 0.6μ, respectively (Harder, 1974). The increase of stalk length at decreasing growth rates may well be a means of increasing the surface to volume ratio, which favours the capture of molecules present in very low concentration. In this respect it is of interest to notice that organisms growing relatively fast at extremely low concentrations of growth-limiting substrates have a relatively high surface to volume ratio (Kuenen *et al.*, 1976; Veldkamp, 1976).

In summary, all bacteria studied thus far show an increase in volume with increasing growth rate. This phenomenon is independent of the kind of substrate limitation. The increase in size appears to be correlated to a change in cell composition, as will be discussed below.

In cultures growing at very low rate, long pleomorphic cells are commonly found. An example was given by Tempest *et al.* (1967) for *Aerobacter aerogenes* at growth rates $< 0.04\mu_{max}$.

A decrease in mean cell volume with decreasing growth rates was also observed in the ciliated protozoon *Tetrahymena pyriformis* (Curds and Cockburn, 1971). The mean organism volume reached a minimum value at a dilution rate of 0.08 h^{-1} and this steady state volume remained constant at dilution rates below 0.08 h^{-1}. It would be of interest to study the fine structure of these slow-growing cells, containing the absolute minimum of essential components.

The profound influence which growth rate may have on cell morphology clearly has consequences for taxonomical studies and should be taken into account in species descriptions.

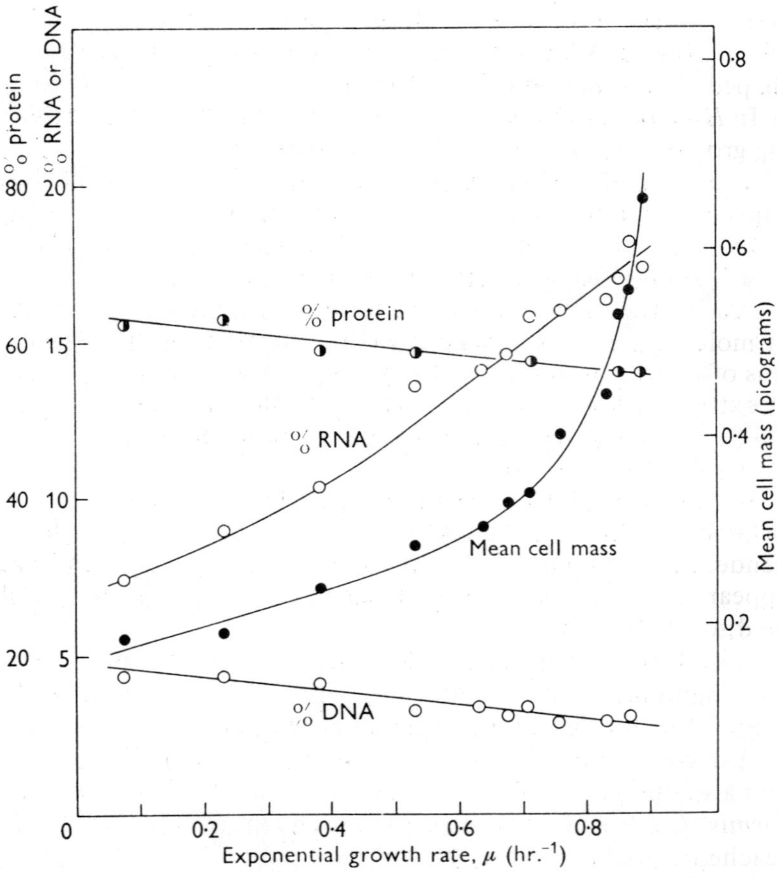

Fig. 18. Growth of *Aerobacter aerogenes* in continuous culture; protein and nucleic acid contents and mean cell mass as functions of growth rate. Nucleic acid and protein contents expressed as percentage of cell dry weight; mean cell mass = dry weight/ml divided by total count/ml. Culture medium: glycerol-NH_3-salts, with glycerol as limiting factor. (Herbert, 1961).

5.2 *Cell Composition*

The most remarkable change in composition with increasing growth rates is the increase in the RNA-percentage of cell dry weight. An example is given in Fig.18.

As described by Schaechter (1973) (see also Maaløe and Kjeldgaard, 1966) for *Salmonella typhimurium*, the number of nuclei per cell increases with growth rate. The increase of cell mass per nucleus with increasing growth rates is due to the larger content of ribosomes in faster-growing cells. At relatively high growth rates and constant temperature the rate of protein synthesis is proportional to the number of ribosomes; thus, to harbour the high number of ribosomes needed at high growth rates, fast-growing cells should be larger than slow-growing cells.

Tempest and coworkers showed in a series of experiments with *Aerobacter aerogenes*, reviewed by Tempest and Dicks (1967) and Tempest (1969) that an increase in RNA/g cells with increasing growth rate is accompanied by an increase in K^+ and Mg^{2+} content. A stoichiometric relation was found between the contents of Mg, K, P, and RNA at all growth rates in cells grown under various nutrient limitations. An example obtained with K-limited cells is shown in Fig.19. The molar RNA/Mg^{2+} ratio shows the highest degree of growth-rate dependence, but this variation is small compared with the total change in RNA content in K-limited cells (8.5 to 18% of the bacterial dry weight, corresponding to growth rates 0.1 to 0.8 h^{-1}).

Slight deviations in the stoichiometry were found in P-limited cells at low growth rates. These were related to increased glycogen synthesis under these conditions (Dicks and Tempest, 1967).

As was discussed by Tempest and Dicks (1967), Mg^{2+} and K^+ play a role in the proper functioning of ribosomes. Thus not only the ribosome content but also the cellular Mg^{2+} and K^+ content is a function of growth rate.

The increase in Mg, K and P per g of cell dry weight with increasing growth rate has an effect on growth yield when these elements limit growth rate. Under these conditions the contents of the elements approximately doubles with an increase in growth rate from 0.10 to $0.68h^{-1}$ in *A. aerogenes* and at the same time the yield decreases by a factor two (Table 2). This is another example of a deviation from theory (eq. 7; Fig. 3), which is illustrated in Fig. 20.

In experiments of this type care must be taken not to change from one growth-limitation to another. When starting with Mg^{2+}-limited cells at low dilution rate, and only a small excess of K^+, then with increasing dilution rates the culture may become K^+-limited (Dicks and Tempest, 1966).

The above examples refer to cell composition as a function of growth rate. A typical example of a specific effect of the kind of growth-limitation was described by Ellwood and Tempest (1972). When Gram-positive organisms are grown with an adequate supply of phosphate, their walls contain a teichoic acid. Wall-bound teichoic acids are polymers of either ribitol phosphate or glycerol phosphate to which sugars and alanine are attached in varying amounts.

In Gram-positive organisms, when grown under P-limitation, wall-teichoic acid was totally replaced by the non-phosphorus-containing anionic polymer, teichuronic acid.

Another example of a specific effect of the kind of growth-limitation is that of diatoms limited by silicate. During blooms in natural environments the silicate concentration could become

Fig. 19. Stoichiometry between cellular RNA and the potassium, magnesium and phosphorus contents of *Aerobacter aerogenes* grown at different dilution rates. The values are expressed as moles RNA-nucleotide (assuming an average molecular weight of 340) per mole potassium, magnesium and phosphorus, respectively. (Tempest *et al.*, 1966; Tempest and Dicks, 1967).

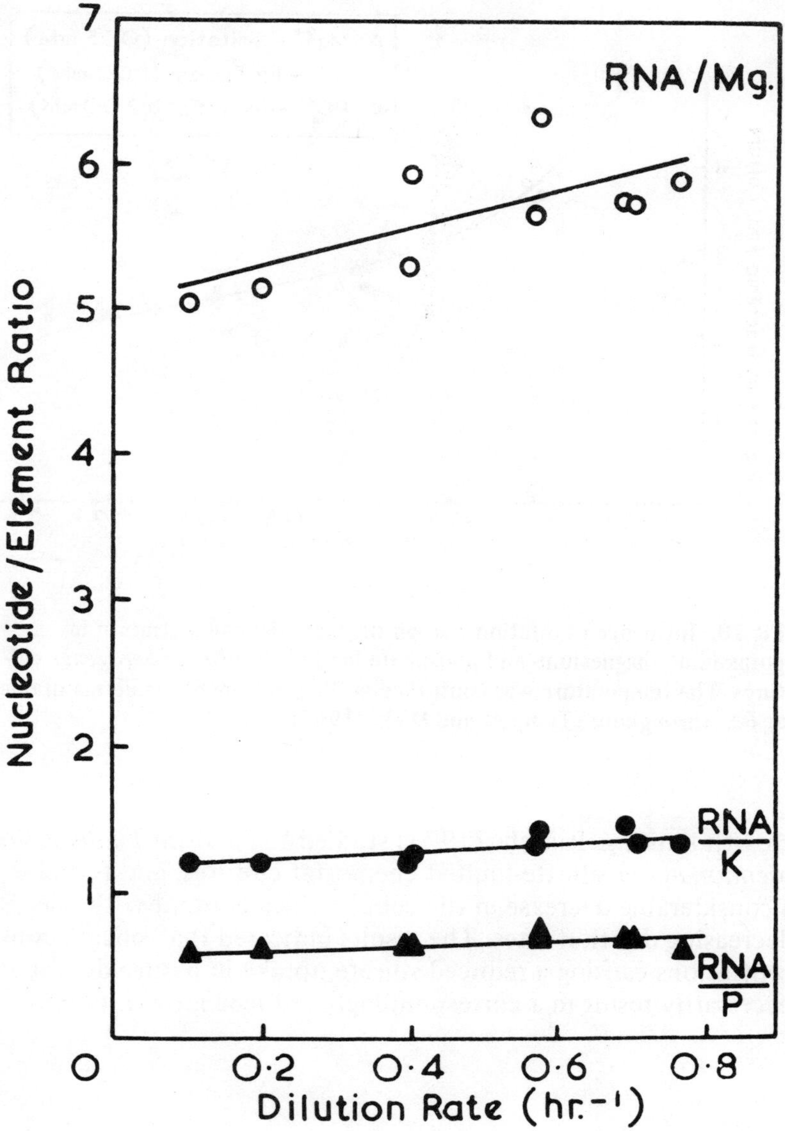

RNA / Mg.
RNA
K
RNA
P
Nucleotide / Element Ratio
7
6
5
4
3
2
1
O
0·2
0·4
0·6
0·8
Dilution Rate (hr.$^{-1}$)

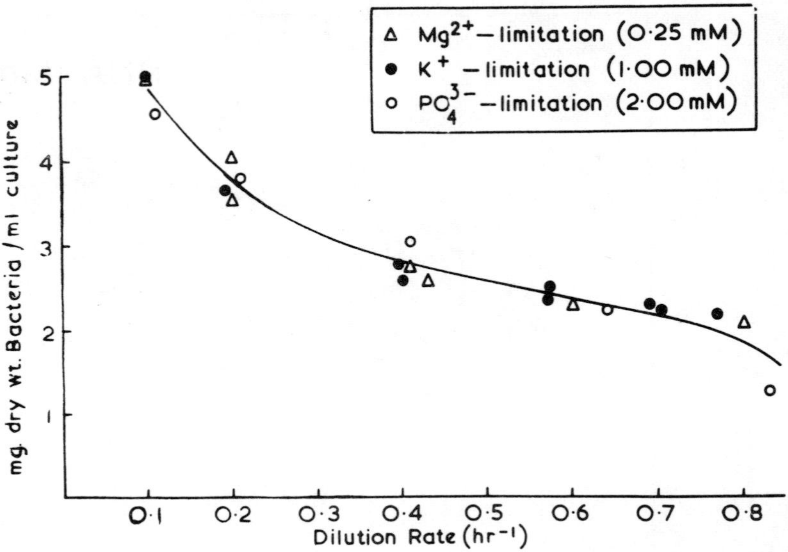

Fig.20. Influence of dilution rate on the bacterial concentration in potassium-, magnesium- and phosphate-limited *Aerobacter aerogenes* cultures. The temperature was controlled at 35°, and the pH value maintained at 6.5, throughout (Tempest and Dicks, 1967).

growth-limiting. Paasche (1973) studied the diatom *Thalissiosira pseudonana* in silicate-limited chemostat cultures, and found a considerable decrease in the cellular silica content with decreasing dilution rates. The results indicated that silicate concentrations causing a reduced silicate uptake in nature do not necessarily result in a correspondingly reduced growth rate.

TABLE 2

Analysis of Mg^{2+}–limited *Aerobacter aerogenes* NCTC 418 grown at different dilution rates (Dicks and Tempest, 1966).

(Values quoted in this table were obtained from samples collected during steady-state growth at each dilution rate (D).)

D h^{-1}	bacterial concentration (equiv. mg dry weight bacteria/ml culture	K^+	Mg^{2+}	P	RNA
			g/100 g dry bacteria		
0.10	5.56	0.89	0.115	1.60	9.7
0.20	3.96	1.14	0.136	1.74	11.1
0.29	3.04	0.14	0.158	1.89	12.1
0.48	2.67	1.52	0.192	2.28	15.4
0.68	2.44	1.67	0.201	2.34	16.8

REFERENCES*

Abson, J.W. and Todhunter, K.H. (1961), in *Continuous Culture of Micro-organisms*, Soc. Chem. Ind., London, Monograph, No.12, 147.

Andrews, J.F. (1968), *Biotechnol. Bioeng.*, **10**, 707.

Baker, K. (1968), *Lab. Pract.*, **17**, 817.

Baschnagel-DePamphilis, J. and Hanson, R.S. (1969), *J. Bacteriol.*, **98**, 222.

Belaich, J.P., Belaich, A. and Simonpietri (1972), *J. gen. Microbiol.*, **70**, 179.

Bongers, L. (1970), *J. Bacteriol.*, **104**, 145.

Bungay, III, H.R. and Bungay, M.L. (1968), *Adv. Appl. Microbiol.*, **10**, 269.

Canale, R.P. (1969), *Biotechnol. Bioeng.*, **11**, 887.

Canale, R.P. (1970), *Biotechnol. Bioeng.*, **12**, 353.

Clarke, P.H. and Lilly, M.D. (1969), in *Microbial Growth*. 19th *Symp. Soc. Gen. Micr.* (Eds. P.M. Meadow and S.J. Pirt). Cambridge University Press, p.113.

Curds, C.R. (1971a), *Water Research*, **5**, 793.

Curds, C.R. (1971b), *Water Research*, **5**, 1049.

Curds, C.R. and Cockburn, A. (1968), *J. gen. Microbiol.*, **54**, 343.

Curds, C.R. and Cockburn, A. (1971), *J. gen. Microbiol.*, **66**, 95.

Curds, C.R., Cockburn, A. and Vandyke, J.M. (1968), *Water Pollut. Control.*, **67**, 312.

Dalton, H. and Postgate, J.R. (1968), *J. gen. Microbiol.*, **54**, 463.

Dalton, H. and Postgate, J.R. (1969), *J. gen. Microbiol.*, **56**, 307.

Dawes, E.A. and Senior, P.J. (1973), *Adv. Microb. Physiol.*, **10**, 135.

Dawson, P.S.S. (1972), *J. appl. Chem. Biotechnol.*, **22**, 79. Also published in *Environmental control of cell synthesis and function* (Eds. A.C.R. Dean, S.J. Pirt and D.W. Tempest), London and New York: Academic Press.

Dean, A.C.R. (1972), *J. appl. Chem. Biotechnol.*, **22**, 245.

Dicks, J.W. and Tempest, D.W. (1966), *J. gen. Microbiol.*, **45**, 547-557.

Dicks, J.W. and Tempest, D.W. (1967), *Biochim. Biophys. Acta*, **136**, 176.

Drozd, J.W., Tubb, R.S. and Postgate, J.R. (1972), *J. gen. Microbiol.*, **73**, 221.

Ellwood, D.C. and Tempest, D.W. (1972), *Adv. Microbial Physiol.*, **7**, 83.
Evans, C.G.T., Herbert, D. and Tempest, D.W., (1970), in *Methods in Microbiology*, vol.2, (Eds. J.R. Norris and D.W. Ribbons), London: Academic Press, p.277.
Fencl, Z. (1966), in *Theoretical and Methodological basis of continuous culture of microorganisms*. (Eds. I. Málek and Z. Fencl). New York and London: Academic Press, p.67.
Fencl, Z., Řičica, J. and Kodesová, J. (1972), *J. appl. Chem. Biotechnol.*, **22**, 405.
Forrest, W.W. and Walker, D.J. (1971), *Adv. Micr. Physiol.*, **5**, 213.
Gaudy, Jr., A.F. and Gaudy, E.T. (1972), *Adv. Biochem. Eng.*, **2**, 97.
Gemerden H. van (1974), *Micr. Ecol.*, **1**, 104.
Gemerden, H. van, and Jannasch, H.W. (1971), *Arch. Mikrobiol.*, **79**, 345.
Haddock, B.A. and Garland, P.B. (1971), *Biochem. J.*, **125**, 155.
Hansford, G.S. and Humphrey, A.E. (1966), *Biotechnol. Bioeng.*, **8**, 85.
Harder, W. (1974), unpublished.
Harder, W. (1976), in 6th International Symposium of Continuous Culture, *"Application and New Fields"* (Eds. A.C.R. Dean, D.C. Ellwood, C.G.T. Evans and J. Melling), Ellis-Horwood Ltd., Chichester.
Harder, W., van Dijken, J.P. (1976), in *Proc. Sympos. Microbial production and utilization of gases.* E. Goltze KG, Göttingen, Germany.
Harder, W. and Veldkamp, H. (1967), *Arch. Mikrobiol.*, **59**, 123.
Harder, W. and Veldkamp, H. (1969), in *Continuous cultivation of microorganisms* (Eds. I. Málek, K. Beran, Z. Fencl, V. Munk, J. Řičica, J. Smrčková), Prague: Academia Publ. House, p.59.
Harder, W. and Veldkamp, H. (1971), *Antonie van Leeuwenhoek*, **37**, 51.
Harder, W., Visser, K. and Kuenen, J.G. (1974), *Lab. Pract.*, **23**, 644.
Harrison, D.E.F. (1973), *Crit. Rev. Microbiol.*, **2**, 185.
Harrison, D.E.F. and Loveless, J.E. (1971), *J. gen. Microbiol.*, **68**, 35.
Harrison, D.E.F., Wilkinson, T., Wren, S. and Harwood, J. (1976), in 6th International Symposium of Continuous Culture *"Application and New Fields"* (Eds. A.C.R. Dean, D.C. Ellwood, C.G.T. Evans and J. Melling), Ellis-Horwood Ltd., Chichester.
Hartley, B.S. (1974), in *Evolution in the microbial world*, 24th *Symp. Soc. Gen. Micr.*, (Eds. M.J. Carlile and J.J. Skehel), Cambridge University Press, p.151.

Hartley, B.S., Burleigh, B.D., Midwinter, C.G., Moore, C.H., Morris, H.R., Rigby, P.W.J., Smith, M.J. and Taylor, S.S. (1972), in *Enzymes: Structure and Function*, 8th FEBS Meeting, vol.29, eds. J. Drenth, R.A. Oosterbaan, C. Veeger, p.151, Amsterdam: North Holland Publ. Co.

Harvey, R.J. (1970), *J. Bacteriol.*, **104**, 698.

Herbert, D. (1958), in *Recent Progress in Microbiology* (Ed. G. Tunevall), 7th Int. Congr. Microbiology, p.381, Stockholm: Almqvist and Wiksell.

Herbert, D. (1961), in *Microbial reaction to environment*. 11th *Symp. Soc. Gen. Micr.* (Eds. G.G. Meynell and H. Gooder). Cambridge University Press, p.391.

Herbert, D. (1964), in *Continuous cultivation of microorganisms*. (Eds. I. Málek, K. Beran and J. Hospodka). New York and London: Academic Press, p.23.

Herbert, D., Elsworth, R. and Telling, R.C. (1956), *J. gen. Microbiol.*, **14**, 601.

Herbert, D., Phipps, P.J. and Tempest, D.W. (1965), *Lab. Pract.*, **14**, 1150.

Hill, S., Drozd, J.W. and Postgate, J.R. (1972), *J. appl. Chem. Biotechnol.*, **22**, 541. Also published in *Environmental control of cell synthesis and function* (Eds. A.C.R. Dean, S.J. Pirt and D.W. Tempest), London and New York: Academic Press.

Hunter, K. and Rose, A.H. (1972), *J. appl. Chem. Biotechnol.*, **22**, 527. Also published in *Environmental control of cell synthesis and function* (Eds. A.C.R. Dean, S.J. Pirt and D.W. Tempest), London and New York: Academic Press.

Jannasch, H.W. (1967), *Arch. Mikrobiol.*, **59**, 165.

Jannasch, H.W. (1974), *Limnol. Oceanogr.*, **19**, 716.

Jannasch, H.W. and Mateles, R.I. (1974), *Adv. Microb. Physiol.*, **11**, 165.

Jensen, D.E. and Neidhardt, F.C. (1969), *J. Bacteriol.*, **98**, 131.

Jost, J.L., Drake, J.F., Fredrickson, A.G. and Tsuchyia, H.M. (1973a), *J. Bacteriol.*, **113**, 834.

Jost, J.L., Drake, J.F., Tsuchiya, H.M. and Fredrickson, A.G. (1973b), *J. theor. Biol.*, **41**, 461.

King, W.R., Sinclair, C.G. and Topiwala, H.H. (1972), *J. gen. Microbiol.*, **71**, 87.

Koch, A.L. (1971), *Adv. Microb. Physiol.*, **6**, 147.

Koch, A.L. and Deppe, C.S. (1971), *J. Mol. Biol.*, **55**, 549.

Kubitschek, H.E. (1970), *Introduction to research with continuous cultures*. Englewood Cliffs: Prentice-Hall.

Kuenen, J.G., Boonstra, J., Schröder, H.G.J. and Veldkamp, H. (1976), *Microbial Ecology*, in preparation.
Lazdunski, A. and Belaich, J.P. (1972), *J. gen. Microbiol.*, **70**, 187.
Light, P.A. (1972), *J. appl. Chem. Biotechnol.*, **22**, 509. Also published in *Environmental control of cell synthesis and function.* (Eds. A.C.R. Dean, S.J. Pirt and D.W. Tempest), London and New York: Academic Press.
Lotka, A.J. (1925), *Elements of physical biology*. Baltimore: Williams and Wilkins.
Luscombe, B.M. and Gray, T.R.G. (1971), *J. gen. Microbiol.*, **69**, 433.
Maaløe, O. (1960), in *Microbial Genetics*, 10th *Symp. Soc. Gen. Microbiol.* (Eds. W. Hayes and R.C. Clowes). Cambridge University Press, p.272.
Maaløe, O. and Kjeldgaard, N.O. (1966), *Control of Macromolecular Synthesis*. New York: Benjamin.
MacLennan, D.G., Pirt, S.J. (1970), *J. Appl. Bacteriol.*, **33**, 390.
Magasanik, B. (1961), *Cold Spring Harbor Symp. Quant. Biol.*, **26**, 249.
Malcolm, N.L. (1968a), *J. Bacteriol.*, **95**, 1388.
Malcolm, N.L. (1968b), *Biochim. Biophys. Acta*, **157**, 493.
Málek, I. and Fencl, Z. (Eds.) (1966), *Theoretical and methodological basis of continuous culture of microorganisms*. New York and London: Academic Press.
Mateles, R.I., Chian, S.K. and Silver, R. (1967), in *Microbial Physiology and Continuous Culture* (Eds. E.O. Powell, C.G.T. Evans, R.E. Strange and D.W. Tempest), London: Her Majesty's Stationery Office, p.232.
Mateles, R.I., Ryu, D.Y. and Yasuda, T. (1965), *Nature*, **208**, 263.
Meers, J.L. (1973), *Crit. Rev. Microbiol.*, **2**, 139.
Meers, J.L. and Tempest, D.W. (1968), *J. gen. Microbiol.*, **52**, 309.
Meers, J.L. and Tempest, D.W. (1970), *Biochem. J.*, **119**, 603.
Meers. J.L., Tempest, D.W. and Brown, C.M. (1970), *J. gen. Microbiol.*, **64**, 187.
Monod, J. (1942), *Recherches sur la croissance des cultures bacteriénnes*, Paris: Hermann.
Monod, J. (1950), *Annls Inst. Pasteur, Paris*, **79**, 390.
Moser, H. (1958), *The Dynamics of bacterial Populations maintained in the Chemostat*. Carnegie Institution Publication No.614, Washington.
Neidhardt, F.C. (1963), *Ann. Rev. Microbiol.*, **17**, 61.
Neidhardt, F.C. and Magasanik, B. (1956), *Biochim. Biophys. Acta*, **21**, 324.
Novick, A. and Szilard, L. (1950), *Proc. nat. Acad. Sci., USA*, **36**, 708.
Paasche, E. (1973), *Mar. Biol.*, **19**, 117.

Payne, W.J. (1970), *Ann. Rev. Microbiol.*, **24**, 17.
Pike, E.B. and Curds, C.R. (1971), in *Microbial Aspects of Pollution*, (Eds. G. Sykes and F.A. Skinner). New York and London: Academic Press.
Pirt, S.J. (1965), *Proc. Roy. Soc. B.*, **163**, 224.
Pirt, S.J. (1975), *Principles of microbe and cell cultivation*. Blackwell Scientific Publications, Oxford, London, Edinburgh, Melbourne.
Postgate, J.R. (1967), *Adv. Micr. Physiol.*, **1**, 1.
Postgate, J.R. (1972), *Biological Nitrogen Fixation*. Newcastle-upon-Tyne, England: Merrow.
Postgate, J.R. (1973), in *Modern methods in the study of microbial ecology* (Ed. T. Rosswall). *Bull. Ecol. Res. Comm. (Stockholm)*, **17**, 287.
Postgate, J.R. and Hunter, J.R. (1962), *J. gen. Microbiol.*, **29**, 233.
Postgate, J.R. and Hunter, J.R. (1963), *J. appl. Bact.*, **26**, 295.
Powell, E.O. (1958), *J. gen. Microbiol.*, **18**, 259.
Powell, E.O. (1965), *Lab. Pract.*, **14**, 1145.
Powell, E.O. (1967), in *Microbial Physiology and Continuous Culture* (Eds. E.O. Powell, C.G.T. Evans, R.E. Strange and D.W. Tempest), p.34. Her Majesty's Stationery Office.
Powell, E.O. (1969), in *Continuous cultivation of microorganisms* (Eds. I. Málek, K. Beran, Z. Fencl, V. Munk, J. Řičica and J. Smrčková), Prague: Academia, p.275.
Powell, E.O. and Lowe, J.R. (1964), in *Continuous cultivation of microorganisms.* (Eds. I. Málek, K. Beran and J. Hospodka). New York and London: Academic Press, p.23.
Řičica, J. (1966), in *Theoretical and methodological basis of continuous culture of microorganisms.* (Eds. I. Málek and Z. Fencl). New York and London: Academic Press, p.155.
Řičica, J. (1970), in *Methods in Microbiology*, vol.2, (Eds. J.R. Norris and D.W. Ribbons), London: Academic Press, p.329.
Řičica, J., Necinová, S., Stejskalová, E. and Fencl, Z. (1967) in *Microbial Physiology and Continuous Culture* (Eds. E.O. Powell, C.G.T. Evans, R.E. Strange, D.W. Tempest). Her Majesty's Stationery Office, p.196.
Rosenberger, R.F. and Elsden, S.R. (1960), *J. gen. Microbiol.*, **22**, 726.
Schaechter, M. (1973), in *Biochemistry of Bacterial Growth*, 2nd ed. (Eds. J. Mandelstam and K. McQuillen). Oxford and Edinburgh: Blackwell Scientific Publishers, p.137.
Schaechter, M., Maaløe, O. and Kjeldgaard, N.O. (1958), *J. gen. Microbiol.*, **19**, 592.

Schulze, K.L. and Lipe, R.S. (1964), *Arch. Mikrobiol.*, **48**, 1.
Senior, P.J., Beech, G.A., Ritchie, G.A.F. and Dawes, E.A. (1972), *Biochem. J.*, **128**, 1193.
Sieburth, J. McN. (1967), *J. exp. mar. Biol. Ecol.*, **1**, 98.
Silver, R.S. and Mateles, R.I., (1969), *J. Bacteriol.*, **97**, 535.
Sinclair, C.G. and Topiwala, H.H. (1970), *Biotechnol. Bioeng.*, **12**, 1069.
Solomons, G.L. (1972), *J. appl. Chem. Biotechnol.*, **22**, 217. Also published in *Environmental control of cell synthesis and function* (Eds. A.C.R. Dean, S.J. Pirt and D.W. Tempest), London and New York: Academic Press.
Spicer, C.C. (1955), *Biometrics*, **11**, 225.
Standing, C.N., Fredrickson, A.G. and Tsuchiya, H.M. (1972), *Appl. Microbiol.*, **23**, 354.
Stouthamer, A.H. (1973), *Antonie van Leeuwenhoek*, **39**, 545.
Stouthamer, A.H. (1976), *Yield studies in microorganisms.* Merrow Publishing Company, Ponteland, Newcastle-upon-Tyne, England.
Stouthamer, A.H. and Bettenhaussen, C. (1973), *Biochim. Biophys. Acta*, **301**, 53.
Strange, R.E. and Hunter, J.R. (1967), in *Microbial Physiology and Continuous Culture* (Eds. E.O. Powell, C.G.T. Evans, R.E. Strange and D.W. Tempest), p.102, Her Majesty's Stationery Office.
Tempest, D.W. (1969), in *Microbial Growth.* 19th *Symp. Soc. Gen. Micr.* (Eds. P.M. Meadow and S.J. Pirt). Cambridge University Press, p.87.
Tempest, D.W. (1970), in *Methods in Microbiology*, vol.2, (Eds. J.R. Norris and D.W. Ribbons), London: Academic Press, p.259.
Tempest, D.W. and Hunter, J.R. (1965), *J. gen. Microbiol.*, **41**, 267.
Tempest, D.W., Hunter, J.R. and Sykes, J. (1965), *J. gen. Microbiol.*, **39**, 355.
Tempest, D.W., Dicks, J.W. and Hunter, J.R. (1966), *J. gen. Microbiol.*, **45**, 135.
Tempest, D.W. and Dicks, J.W. (1967), in *Microbial Physiology and Continuous Culture* (Eds. E.O. Powell, C.G.T. Evans, R.E. Strange and D.W. Tempest), p.140. Her Majesty's Stationery Office.
Tempest, D.W., Herbert, D. and Phipps, P.J. (1967), in *Microbial Physiology and Continuous Culture* (Eds. E.O. Powell, C.G.T. Evans, R.E. Strange and D.W. Tempest), p.240, Her Majesty's Stationery Office.
Tempest, D.W., Meers, J.L. and Brown, C.M. (1970), *Biochem. J.*, **117**, 405.
Topiwala, H.H. and Hamer, G. (1971), *Biotechnol. Bioeng.*, **13**, 919.

Tsuchiya, H.M., Drake, J.F., Jost, J.L. and Fredrickson, A.G. (1972), *J. Bacteriol.*, **110**, 1147.
Uden, N. van (1967), *Arch. Mikrobiol.*, **58**, 145.
Uden, N. van (1969), *Ann. Rev. Microbiol.*, **23**, 473.
Veldkamp, H. (1975), in *Unifying concepts in ecology*. (Eds. W.H. van Dobben and R.H. Lowe-McConnell), Report of the plenary sessions of the First international congress of ecology, The Hague, The Netherlands. The Hague: W. Junk Publishers and Wageningen: Pudoc, p.44.
Veldkamp, H. (1976), in 6th International Symposium of Continuous Culture *"Application and New Fields"*, (Eds. A.C.R. Dean, D.C. Ellwood, C.G.T. Evans and J. Melling), Ellis-Horwood Ltd., Chichester.
Veldkamp, H. and Jannasch, H.W. (1972), *J. appl. Chem. Biotechnol.*, **22**, 105. Also published in *Environmental control of cell synthesis and function* (Eds. A.C.R. Dean, S.J. Pirt and D.W. Tempest), London and New York: Academic Press.
Veldkamp, H. and Kuenen, J.G. (1973), in *Modern methods in the study of microbial ecology* (Ed. T. Rosswall). *Bull. Ecol. Res. Comm. (Stockholm)*, **17**, 347.
Volterra, V. (1926), *Mem. Acad. Lineii Roma*, **2**, 31.
Vries, W. de, Kapteijn, W.M.C., van der Beek, E.G. and Stouthamer, A.H. (1970), *J. gen. Microbiol.*, **63**, 333.
Watson, T.G. (1970), *J. gen. Microbiol.*, **64**, 91.
Watson, T.G. (1972), *J. appl. Chem. Biotechnol.*, **22**, 229. Also published in *Environmental control of cell synthesis and function* (Eds. A.C.R. Dean, S.J. Pirt and D.W. Tempest), London and New York: Academic Press.
Wilkinson, T.G. and Hamer, G. (1974), *Biotechnol. Bioeng.*, **16**, 251.

* Abstracts of continuous culture literature are given annually in *Folia Microbiologica*.